endorsed by edexcel

A Level Mathematics for Edexcel

Mechanics

M2

Brian Jefferson

OXFORD
UNIVERSITY PRESS

OXFORD
UNIVERSITY PRESS

Great Clarendon Street, Oxford OX2 6DP

Oxford University Press is a department of the University of Oxford.
It furthers the University's objective of excellence in research, scholarship, and education by publishing worldwide in

Oxford New York

Auckland Cape Town Dar es Salaam Hong Kong Karachi
Kuala Lumpur Madrid Melbourne Mexico City Nairobi
New Delhi Shanghai Taipei Toronto

With offices in

Argentina Austria Brazil Chile Czech Republic France Greece
Guatemala Hungary Italy Japan South Korea Poland Portugal
Singapore Switzerland Thailand Turkey Ukraine Vietnam

Oxford is a registered trade mark of Oxford University Press in the UK and in certain other countries

First published 2009

British Library Cataloguing in Publication Data

Data available

ISBN 978-0-19-911786 4
10 9 8 7 6 5 4 3

Printed in Great Britain by Ashford Colour Press Ltd

Paper used in the production of this book is a natural, recyclable product made from wood grown in sustainable forests. The manufacturing process conforms to the environmental regulations of the country of origin.

Acknowledgements

The Photograph on the cover is reproduced courtesy of Getty/Photographers Choice

Series Managing Editor Anna Cox
The Publisher would like to thank the following for permission to reproduce photographs:
P28 Norbert Judkowiak/iStockphoto; **P52** CW Motorsport Images/Alamy;
P84 Photodisc **P100** Richard Olivier/Corbis; **P112** Dave Mcaleavy/Dreamstime.

The Publisher would also like to thank John Rayneau, Kathleen Austin and Charlie Bond for their expert help in compiling this book.

About this book

Endorsed by Edexcel, this book is designed to help you achieve your best possible grade in Edexcel GCE Mathematics Mechanics 2 unit.

Each chapter starts with a list of objectives and a 'Before you start' section to check that you are fully prepared. Chapters are structured into manageable sections, and there are certain features to look out for within each section:

Key points are highlighted in a blue panel.

Key words are highlighted in **bold blue** type.

Worked examples demonstrate the key skills and techniques you need to develop. These are shown in boxes and include prompts to guide you through the solutions.

EXAMPLE 8

A particle starts from O with velocity $(10\mathbf{i} + 15\mathbf{j})\,\text{m s}^{-1}$, where $\mathbf{i}$ and $\mathbf{j}$ correspond to the horizontal and vertically upward directions. Find the distance of the particle from O after 2 s.

Use $\mathbf{r} = \mathbf{u}t + \frac{1}{2}\mathbf{a}t^2$ with $t = 2$:

$$\mathbf{r} = (10\mathbf{i} + 15\mathbf{j}) \times 2 - 4.9\mathbf{j} \times 4 = 20\mathbf{i} + 10.4\mathbf{j}$$

The distance d m of the particle from O is the magnitude of its displacement vector so

$$d = \sqrt{20^2 + 10.4^2} = 22.5$$

The particle is 22.5 m from O.

Derivations and additional information are shown in a panel.

Helpful hints are included as blue margin notes and sometimes as blue type within the main text.

Misconceptions are shown in the right margin to help you avoid making common mistakes.

Investigational hints to prompt you to explore a concept further.

Each section includes an exercise with progressive questions, starting with basic practice and developing in difficulty. Some exercises also include 'stretch and challenge' questions marked with a stretch symbol.

At the end of each chapter there is a 'Review' section which includes exam style questions as well as past exam paper questions. There are also two 'Revision' sections per unit which contain questions spanning a range of topics to give you plenty of realistic exam practice.

The final page of each chapter gives a summary of the key points, fully cross-referenced to aid revision. Also, a 'Links' feature provides an engaging insight into how the mathematics you are studying is relevant to real life.

At the end of the book you will find full solutions, a key word glossary, a list of formulae you will be given in the exam and an index.

Contents

M2

1 Kinematics

This chapter will show you how to
- model the motion of a projectile fired from ground level or from a point above ground level
- calculate the range, time of flight and maximum height of such a particle
- use calculus to find the position, velocity and acceleration of a particle moving with variable acceleration
- differentiate and integrate vectors with respect to time.

Before you start

You should know how to:

1 Manipulate vectors in two dimensions.

Check in:

1

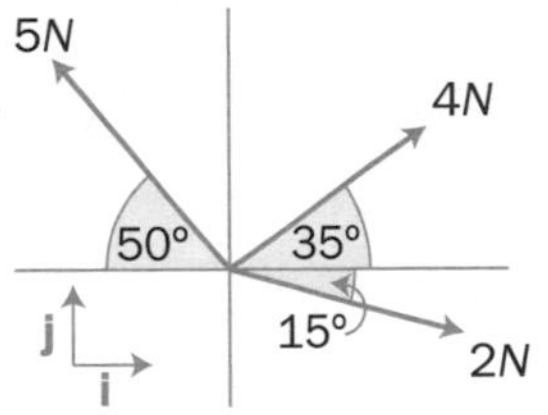

a Find in component form the resultant of the vectors shown.

b Find the magnitude and direction of the resultant.

2 Use the formulae relating to motion with constant acceleration.

2 A car moving in a straight line at $6\,\mathrm{m\,s^{-1}}$ accelerates to a speed of $10\,\mathrm{m\,s^{-1}}$ while travelling 64 m. Find

a its acceleration

b the length of time for which it was accelerating.

3 Differentiate polynomials and algebraic, exponential, logarithmic and trigonometric functions.

3 Find $\frac{dy}{dx}$ where

a $y = 3x^3 + 2x$

b $y = x^{\frac{5}{2}} + \frac{7}{\sqrt{x}} + \frac{3}{x^2}$

c $y = \ln(2x + 1)$

d $y = e^{5x}$

e $y = \sin 4x$

f $y = x^3 \cos 2x$

4 Integrate polynomials and algebraic functions.

4 Find

a $\int (x^4 + 2x^3)\,dx$

b $\int \frac{2x + 5}{\sqrt{x}}\,dx$

M2

1.1 Projectiles

In M1 you modelled the motion of an object moving vertically under gravity. However, if you throw an object into the air, it does not usually move vertically. It has a horizontal and a vertical displacement from its starting point, and horizontal and vertical components of velocity. The motion of a projectile is two-dimensional motion with constant acceleration.

You assume the object is a particle and that there is no air resistance as you did in M1.

For the general 2-D problem you can use the notation

displacement $\mathbf{r} = x\mathbf{i} + y\mathbf{j}$ initial velocity $\mathbf{u} = u_x\mathbf{i} + u_y\mathbf{j}$

final velocity $\mathbf{v} = v_x\mathbf{i} + v_y\mathbf{j}$ acceleration $\mathbf{a} = a_x\mathbf{i} + a_y\mathbf{j}$

a_x and a_y are constant, so the motion in each direction obeys the equations for constant acceleration. So, for example,

$v_x = u_x + a_x t$ and $v_y = u_y + a_y t$

You can write these as a single vector equation

$\mathbf{v} = \mathbf{u} + \mathbf{a}t$

In the same way you have

$$\mathbf{r} = \frac{1}{2}(\mathbf{u} + \mathbf{v})t$$

$$\mathbf{r} = \mathbf{u}t + \frac{1}{2}\mathbf{a}t^2$$

$$\mathbf{r} = \mathbf{v}t - \frac{1}{2}\mathbf{a}t^2$$

Apply the equation $v^2 = u^2 + 2as$ to the separate components:

$$v_x^2 = u_x^2 + 2a_x x \quad \text{and} \quad v_y^2 = u_y^2 + 2a_y y$$

See M1 for revision of the equations of motion.

You can also write $v^2 = u^2 + 2as$ in a vector form but this uses techniques beyond the scope of the M2 unit.

For a projectile you take the horizontal (to the right) and vertical (upward) directions as the positive x and y ($\mathbf{i}$ and $\mathbf{j}$) directions.

The diagram shows the path of a particle projected from an origin O with speed u at an angle α to the horizontal.

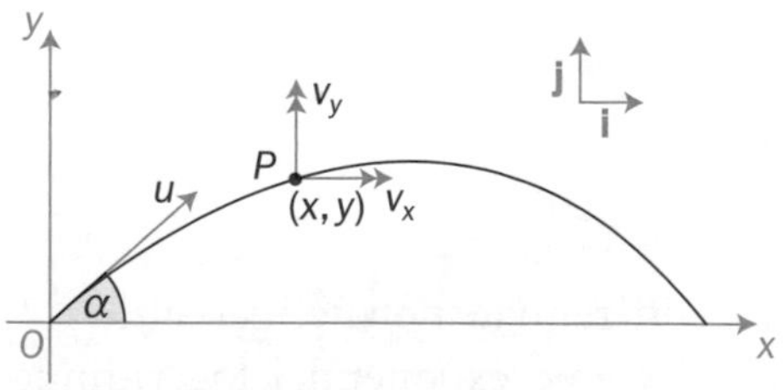

The horizontal and vertical components of its initial velocity are $u_x = u\cos\alpha$ and $u_y = u\sin\alpha$ respectively, so

$$\mathbf{u} = u\cos\alpha\,\mathbf{i} + u\sin\alpha\,\mathbf{j}$$

At a general time, t s, the particle is at point $P(x, y)$.

Its displacement is $\mathbf{r} = x\mathbf{i} + y\mathbf{j}$

Its velocity is $\mathbf{v} = v_x\mathbf{i} + v_y\mathbf{j}$

Its acceleration is $\mathbf{a} = -g\mathbf{j}$

The negative sign is because acceleration due to gravity, g, acts downwards. There is no horizontal acceleration since air resistance is ignored.

EXAMPLE 1

A particle is projected from an origin O at $20\,\text{m s}^{-1}$ at an angle of 35° to the horizontal. Find

a its speed and direction **b** its position

after 2 s.

Initial velocity is $\mathbf{u} = 20\cos 35°\,\mathbf{i} + 20\sin 35°\,\mathbf{j}$

Acceleration is $\mathbf{a} = -9.8\,\mathbf{j}$

a Use $\mathbf{v} = \mathbf{u} + \mathbf{a}t$ with $t = 2$:

$$\begin{aligned}\mathbf{v} &= 20\cos 35°\,\mathbf{i} + 20\sin 35°\,\mathbf{j} - 9.8\,\mathbf{j} \times 2\\ &= 20\cos 35°\,\mathbf{i} + (20\sin 35° - 19.6)\,\mathbf{j}\\ &= 16.4\,\mathbf{i} - 8.13\,\mathbf{j}\end{aligned}$$

The speed of the particle is given by

$$v = \sqrt{16.4^2 + 8.13^2} = 18.3$$

Its direction is given by

$$\tan\theta = \frac{8.13}{16.4} \quad \text{so} \quad \theta = 26.4°$$

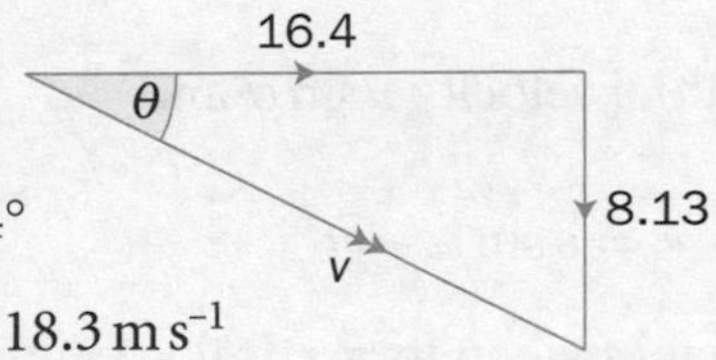

A diagram helps you to visualise the problem.

So the particle is travelling at $18.3\,\text{m s}^{-1}$ at 26.4° below the horizontal.

b Use $\mathbf{r} = \mathbf{u}t + \frac{1}{2}\mathbf{a}t^2$ with $t = 2$:

$$\begin{aligned}\mathbf{r} &= (20\cos 35°\,\mathbf{i} + 20\sin 35°\,\mathbf{j}) \times 2 - 4.9\mathbf{j} \times 4\\ &= (20\cos 35° \times 2)\mathbf{i} + (20\sin 35° \times 2 - 19.6)\mathbf{j}\\ &= 32.8\mathbf{i} + 3.34\mathbf{j}\end{aligned}$$

So the particle is at the point with position vector $(32.8\mathbf{i} + 3.34\mathbf{j})$ m.

It is often convenient to deal separately with the motion in the two directions, as shown in Example 2.

EXAMPLE 2

A ball is placed on horizontal ground 7.2 m from a vertical wall of height 5 m. The ball is kicked towards the wall with initial speed $12\,\text{m s}^{-1}$ at an angle of 60° to the horizontal. Does the ball go over the wall?

You need to find how high the ball is at the moment it reaches the wall. You must first find how long the ball takes to reach the wall, then find its height at that time.

Consider the horizontal motion:

The initial speed is $12\cos 60° = 6\,\text{m s}^{-1}$.

There is no horizontal acceleration, so after t s it has travelled $6t$ m.

It reaches the wall when $6t = 7.2$, giving $t = 1.2$ s.

Example 2 is continued on the next page.

M2

EXAMPLE 2 (CONT.)

Now consider the vertical motion:
The initial speed is $12\sin 60° = 6\sqrt{3}\,\text{m s}^{-1}$.
The vertical acceleration is $-9.8\,\text{m s}^{-2}$.

$\sin 60° = \frac{\sqrt{3}}{2}$

To find the height, y, use $s = ut + \frac{1}{2}at^2$ with $t = 1.2$:

$$y = 6\sqrt{3} \times 1.2 - 4.9 \times 1.2^2 = 5.41$$

The wall is 5 m high, so the ball clears it by 0.41 m.

You may need to find the **maximum height** reached by a projectile.

A projectile leaves the origin O with speed $u\,\text{m s}^{-1}$ and elevation α.

In the vertical direction it has initial velocity $u\sin\alpha$ and acceleration $-g\,\text{m s}^{-2}$.
At time t s, its vertical velocity is $v_y = u\sin\alpha - gt$

The projectile is at maximum height when its vertical velocity becomes zero, so

$$u\sin\alpha - gt = 0 \quad \text{giving} \quad t = \frac{u\sin\alpha}{g}$$

Use $s = ut + \frac{1}{2}at^2$: Maximum height, H m, is given by

$$H = u\sin\alpha \times \frac{u\sin\alpha}{g} - \frac{1}{2}g \times \frac{u^2\sin^2\alpha}{g^2}$$

$$= \frac{u^2\sin^2\alpha}{2g}$$

This gives a formula for maximum height. In the examination you will be expected to derive your result from the basic equations, as in Example 3. It is not sufficient to just quote the formula.

EXAMPLE 3

A golf ball resting on horizontal ground is struck so that it initially moves at $50\,\text{m s}^{-1}$ at an angle of 20° to the horizontal. Find the greatest height to which it rises.

Consider the vertical motion:
Initial velocity is $50\sin 20° = 17.1\,\text{m s}^{-1}$.
Acceleration is $-9.8\,\text{m s}^{-2}$.
At time ts, vertical velocity is $v_y = 17.1 - 9.8t$
At maximum height $v_y = 0$, so $17.1 - 9.8t = 0$ giving $t = 1.745$ s.

Rounded values are shown here. You should not round during your working as this may affect the accuracy of your answer.

To find maximum height, H, use $s = ut + \frac{1}{2}at^2$ with $t = 1.745$:

$$H = 17.1 \times 1.745 - 4.9 \times 1.745^2 = 14.9$$

So the maximum height reached by the ball is 14.9 m.

You may need to find the range of a projectile, that is the distance it travels before hitting the ground.

You can assume that the ground is horizontal.

A projectile leaves the origin O with speed $u\,\text{m s}^{-1}$ and elevation α.

In the vertical direction it has initial velocity $u\sin\alpha$ and acceleration $-g\,\text{m s}^{-2}$.

Use $s = ut + \frac{1}{2}at^2$: The height of the projectile at time t s is

$$y = ut\sin\alpha - \frac{1}{2}gt^2$$

The projectile hits the ground when the height becomes zero, so

$$ut\sin\alpha - \frac{1}{2}gt^2 = 0 \quad \text{giving} \quad t = \frac{2u\sin\alpha}{g} \quad (\text{or } t = 0)$$

The root $t = 0$ occurs because the height was zero at the start.

The time for which the projectile is in the air is the time of flight.

The time of flight is twice the time taken to reach maximum height. Unless there is air resistance, the motion of the projectile is symmetrical.

To find the range, consider the horizontal motion:

The horizontal velocity component is $u\cos\alpha$, and this is constant.

Range R m is given by

$$R = u\cos\alpha \times \frac{2u\sin\alpha}{g} = \frac{2u^2\sin\alpha\cos\alpha}{g}$$

You can write this as $R = \dfrac{2u^2\sin\alpha\sin(90° - \alpha)}{g}$

This gives a formula for the range. In the examination you will be expected to derive your result from the basic equations, as in Example 4. It is not sufficient to just quote the formula.

This means that for a given initial speed you get the same range if the angle of projection is α or $(90° - \alpha)$. The greatest possible range occurs when $\alpha = (90° - \alpha)$, that is when $\alpha = 45°$.

The formula can be written as $R = \dfrac{u^2\sin 2\alpha}{g}$, so R is a maximum when $\sin 2\alpha = 1$, giving $\alpha = 45°$

M2

EXAMPLE 4

A golf ball is hit towards a hole 180 m away with velocity $50\,\text{m s}^{-1}$ at an angle of 30° to the horizontal. Assuming horizontal ground, how far from the hole will the ball land?

The initial vertical velocity component is

$$u_y = 50\sin 30° = 25\,\text{m s}^{-1}$$

The vertical displacement of the ball at time t is

$$y = 25t - 4.9t^2$$

Using $s = ut + \frac{1}{2}at^2$

The ball hits the ground when $y = 0$, so

$$25t - 4.9t^2 = 0$$
$$t(25 - 4.9t) = 0$$
$$t = 0 \quad \text{or} \quad t = 5.1$$

The time of flight is 5.1 s.

Example 4 is continued on the next page.

EXAMPLE 4 (CONT.)

The horizontal velocity component is $u_x = 50\cos 30° = 25\sqrt{3}$, and this is constant.

$\cos 30° = \frac{\sqrt{3}}{2}$

The horizontal displacement of the ball at time t is $x = 25\sqrt{3}t$.

Substitute $t = 5.1$:

Range is $25\sqrt{3} \times 5.1 = 220.9$

The hole is 180 m away, so the ball lands 40.9 m beyond the hole.

$220.9 - 180 = 40.9$

Examples 5, 6 and 7 explore some other situations that you might meet.

EXAMPLE 5

Sunil throws a stone horizontally from the top of a 40 m cliff at a speed which is just sufficient to reach the water's edge 50 m from the base of the cliff. Calculate

a the initial speed of the stone

b the speed and direction of the stone as it hits the water.

---

a Use $s = ut + \frac{1}{2}at^2$:

The initial vertical velocity is zero.

Vertical displacement at time t is $y = -4.9t^2$.

When the stone hits the water $y = -40$ so

$-4.9t^2 = -40$ giving $t = 2.86$

Sketching diagrams will help you to visualise the problem.

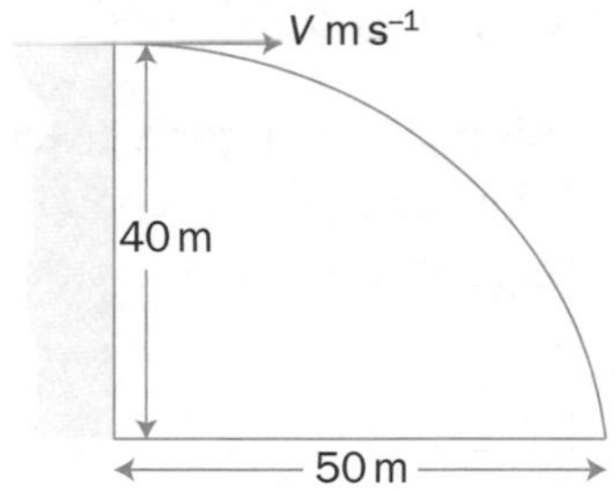

The stone must travel 50 m horizontally in 2.86 s.
The horizontal velocity is $V\,\text{m s}^{-1}$ and is constant, so

$$V = \frac{50}{2.86} = 17.5$$

The initial speed of the stone is $17.5\,\text{m s}^{-1}$.

b When the stone reaches the water, its horizontal velocity component is $17.5\,\text{m s}^{-1}$.

Use $v = u + at$:

Its vertical velocity component is

$v_y = -9.8 \times 2.86 = -28$

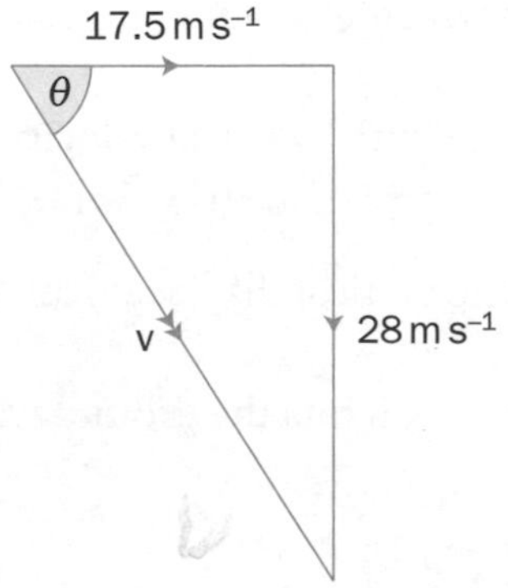

From the diagram

$$v = \sqrt{17.5^2 + 28^2} = 33$$

and $\tan\theta = \frac{28}{17.5}$ so $\theta = 58°$

The stone hits the water at $33\,\text{m s}^{-1}$ at an angle of 58° below the horizontal.

EXAMPLE 6

A particle starts from O with velocity $(10\mathbf{i} + 15\mathbf{j})$ m s^{-1}, where **i** and **j** correspond to the horizontal and vertically upward directions. Find the distance of the particle from O after 2 s.

Use $\mathbf{r} = \mathbf{u}t + \frac{1}{2}\mathbf{a}t^2$ with $t = 2$:

$$\mathbf{r} = (10\mathbf{i} + 15\mathbf{j}) \times 2 - 4.9\mathbf{j} \times 4 = 20\mathbf{i} + 10.4\mathbf{j}$$

The distance d m of the particle from O is the magnitude of its displacement vector so

$$d = \sqrt{20^2 + 10.4^2} = 22.5$$

The particle is 22.5 m from O.

Rather than using the vector form of the constant acceleration equation, you could work with the **i** and **j** components separately.

You may find it useful to sketch a diagram. A triangle of displacements may help.

EXAMPLE 7

A particle is projected at 45 m s^{-1} from a point O on a horizontal plane. The maximum height it reaches is 40 m. Calculate

a the angle α between its initial direction and the horizontal

b the range of the particle.

a The initial vertical component of velocity is $45 \sin \alpha$. At maximum height the vertical velocity component is zero.

Use $v^2 = u^2 + 2as$:

$$0 = (45 \sin \alpha)^2 - 2 \times 9.8 \times 40$$

$$\sin \alpha = \sqrt{\frac{784}{2025}} = 0.622$$

$$\alpha = 38.5°$$

b Find the height, y, of the particle at time t s by using $s = ut + \frac{1}{2}at^2$:

$$y = 45t \sin 38.5° - 4.9t^2$$

The particle hits the plane when $y = 0$, so

$$0 = 45t \sin 38.5° - 4.9t^2$$

$$t(45 \sin 38.5° - 4.9t) = 0$$

$$t = 5.71 \text{ (or } t = 0)$$

The time of flight is 5.71 s.

The horizontal velocity component is $45 \cos 38.5° = 35.2$ m s^{-1}, and this is constant. When $t = 5.71$ the horizontal displacement is

$$x = 35.2 \times 5.71 = 201.3$$

So the range is 201 m (3 s.f.).

Use the stored, accurate values to calculate the final answer, not the rounded values shown in the working here.

M2

Exercise 1.1

1 A stone is thrown horizontally from the top of a 50 m high vertical cliff at 30 m s^{-1}. Find how far from the foot of the cliff the stone hits the sea.

2 A dart is thrown horizontally at 21 m s^{-1} towards a board 3 m away. The point of projection of the dart is at the same level as the bullseye on the board.

a Find the length of time the dart takes to reach the board.

b Find the distance below the bullseye at which the dart strikes the board.

c At what angle to the horizontal is the dart travelling at the moment that it strikes the board?

3 A projectile is launched from a point O on horizontal ground with speed 15 m s^{-1} at an angle of 35° to the horizontal.

a For how long is the projectile in the air?

b What is the horizontal range of the projectile?

c Find the time taken to reach maximum height.

d What is the greatest height reached?

4 A projectile is launched from a point O on horizontal ground with speed 10 m s^{-1} at an angle of 60° to the horizontal.

a For how long is the projectile in the air?

b What is the horizontal range of the projectile?

c Find the time taken to reach maximum height.

d What is the greatest height reached?

5 A particle is projected from a point O on horizontal ground with speed 24 m s^{-1} at an angle of 50° to the horizontal. There is a wall 30 m from O in the direction in which the particle is heading.

a Find how far up the wall the particle strikes it.

b What is the speed of the particle when it hits the wall?

c Find the direction of motion of the particle when it hits the wall.

6 A ball is projected from a point O on horizontal ground with velocity $(30\mathbf{i} + 40\mathbf{j})$ m s^{-1}. Find

a the position of the ball after

i 1 s   ii 2 s

b the length of time for which the ball is in the air

c the range.

7 A particle is projected at $196\,\text{m s}^{-1}$ at an angle α to the horizontal. It reaches a maximum height of 490 m.
Find the value of α.

8 A particle is projected at 45° to the horizontal from a point O on horizontal ground. It strikes the ground again 100 m away.
Find the speed with which it was projected.

9 A ball is kicked from the floor of a gym at 40° to the horizontal and at a speed of $V\,\text{m s}^{-1}$. The ceiling of the gym is 10 m above the floor.

a Find, in terms of V, the maximum height to which the ball rises.

b Hence find the greatest value of V for which the ball does not hit the ceiling.

c For this value of V, find the distance the ball travels before hitting the floor.

d State the main assumptions you have made.

10 Gina is standing outside a football ground during a match.
She can see the ball if it rises above a height of 10 m.
If the ball is kicked from ground level at an angle of 50° to the horizontal and with speed $25\,\text{m s}^{-1}$, how long is it visible to Gina?

Find the two values of t when the height is 10 m.

11 A particle is projected from a point O on horizontal ground.
After 2 seconds it is 40 m horizontally from O and at a height of 30 m.

a Calculate the horizontal and vertical components of the particle's initial velocity.

b How far from O will the particle strike the ground?

c Calculate the speed and direction of the particle as it hits the ground.

12 (Take **i** and **j** as the horizontal and vertically upward directions.)
A particle is projected with velocity $(u_x\mathbf{i} + u_y\mathbf{j})\,\text{m s}^{-1}$ from a point P with position vector $(10\mathbf{i} + 5\mathbf{j})$ m relative to an origin O.
After 2 seconds it is at the point $(25\mathbf{i} + 55\mathbf{j})$ m.

a Calculate the values of u_x and u_y.

b Calculate the distance of the particle from P after a further 1 second.

13 A tower of height h m stands on horizontal ground. An arrow is fired from the top of the tower with speed $40\,\text{m s}^{-1}$ at an angle of 25° above the horizontal. It hits the ground 170 m from the base of the tower. Calculate the value of h.

14 A crow picks up a piece of cheese from horizontal ground and flies off. At a certain instant the crow is 10 m above the ground and is travelling at 12 m s^{-1} in a direction 30° above the horizontal. At that moment it drops the cheese.

a For how long is the cheese in the air after it is released?

b If the crow continues to fly at the same velocity, what is the distance between the crow and the cheese at the moment that the cheese hits the ground?

The initial velocity of the cheese will be equal to the velocity of the crow at that moment.

15 A footballer is 25 m from goal as shown in the diagram. He tries to kick the ball over the goalkeeper, who is 15 m from his goal. He kicks the ball at 30° to the horizontal at a speed of $u\text{ m s}^{-1}$. He will score provided the ball is more than 3 m above the ground when it reaches the goalkeeper and less than 3 m above the ground when it reaches the goal. Find the range of values of u for which he will score.

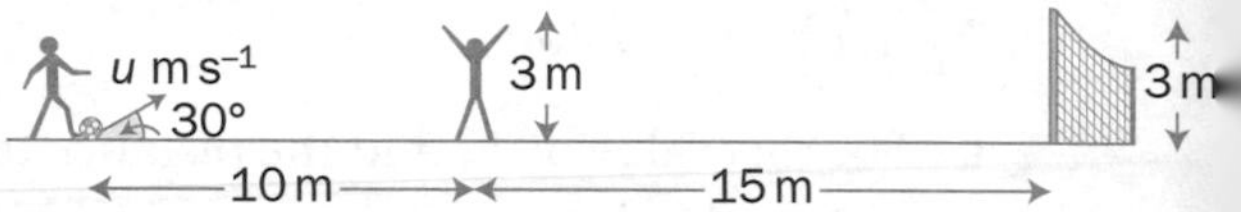

16 A tennis player serves the ball. At the moment that the racket strikes the ball it is 11.6 m from the net and 3.2 m above the ground. The racket imparts a horizontal speed $u\text{ m s}^{-1}$ and a downward vertical speed $v\text{ m s}^{-1}$. The ball hits the ground 6.4 m beyond the net after 0.3 s.

a Calculate the values of u and v.

b By how much does the ball clear the net, if the net is 0.91 m high?

17 A particle is projected from the origin O. It passes horizontally through the point with position vector $(20\mathbf{i} + 10\mathbf{j})$ m. Find the velocity with which it left O.

18 A projectile has a range of 300 m, which it travels in 8 s. Calculate

a the initial speed and direction of the projectile

b the speed and direction after 2 s

c the length of time for which the projectile is at a height greater than 40 m.

19 A particle is projected from an origin O on horizontal ground with initial velocity $(20\mathbf{i} + 30\mathbf{j})\text{ m s}^{-1}$.

a Write down its position vector $\mathbf{r} = x\mathbf{i} + y\mathbf{j}$ at time t.

b By eliminating t from the equations for x and y, find the equation of the path of the particle in the form $y = f(x)$

c Use your answer to part **b** to find

i the value of y when $x = 40$ ii the values of x when $y = 30$

iii the range of the particle.

M2

20 A particle is projected from a point O with speed $V_1\,\mathrm{m\,s^{-1}}$ at an elevation of 60°. One second later another particle is projected from O, in the same vertical plane, with speed $V_2\,\mathrm{m\,s^{-1}}$ at an elevation of 30°. After a further second the particles collide. Find, in terms of g, the values of V_1 and V_2, leaving your answers in surd form.

21 A particle is projected with elevation α from a point O on horizontal ground. It passes through a point with position vector $(p\mathbf{i} + q\mathbf{j})$ and hits the ground at a distance r from O. Find α in terms of p, q and r.

22 A particle projected from a point O has vertical and horizontal velocity components u and v. It passes through the point with position vector $(p\mathbf{i} + q\mathbf{j})$.

a Show that $2qu^2 + gp^2 = 2puv$

A particle is projected so as to pass through points with position vectors $(20\mathbf{i} + 10\mathbf{j})$ and $(50\mathbf{i} + 20\mathbf{j})$.

b Find the speed and elevation with which it was projected.

23 Two particles are projected simultaneously from the same point with the same speed, one with an elevation θ and the other with elevation α. Show that during the flight

a the line joining the two particles has a constant gradient

b the distance between the particles increases at a constant rate.

24 A particle P is projected from a point O with speed V and at an angle θ to the horizontal. The line OQ corresponds to the original direction of the particle, and QP is the vertical line through the particle at time t. The lengths OQ and QP are called the Littlewood coordinates of P and are represented by ξ and η.

Littlewood coordinates are not covered in the Edexcel exam specification for unit M2. You do not need to recall them.

a Show that ξ and η are independent of θ.

A particle is projected with speed V from a point O on a plane inclined at α to the horizontal. Its initial direction makes an angle β with the plane.

b Use Littlewood coordinates, together with the sine rule, to find the time of flight of the particle.

M2

1.2 Motion with non-uniform acceleration

In M1 you used graphs of motion with constant velocity or constant acceleration. In particular you learned that
velocity = gradient of displacement–time (x, t) graph
and
acceleration = gradient of velocity–time (v, t) graph

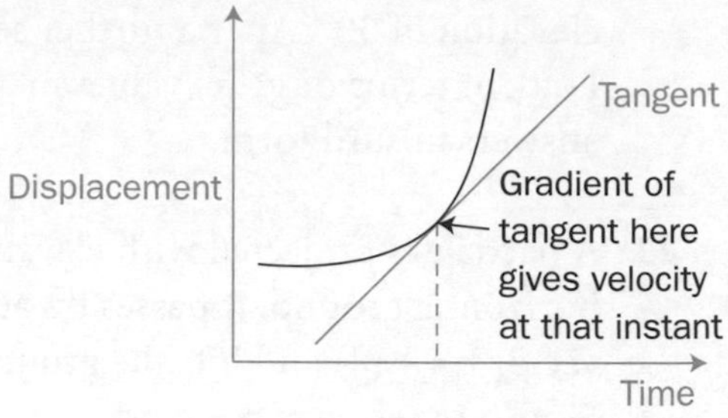

For non-uniform motion the graph is not linear, so the gradient changes, but at a particular point on the graph you still have

gradient of (x, t) graph = velocity at that instant
and gradient of (v, t) graph = acceleration at that instant

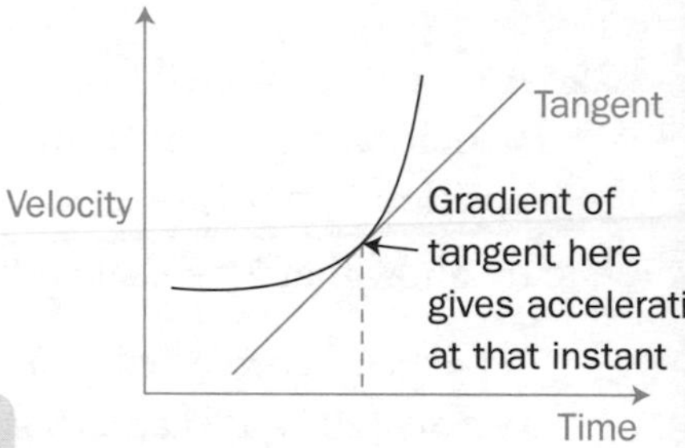

If you know the displacement x as a function of t, you find the gradient (velocity) by differentiation.
The velocity is the rate of change of displacement.

$$v = \frac{dx}{dt}$$

M2

Similarly, acceleration is the rate of change of velocity. You obtain acceleration by differentiating velocity with respect to time, which is the same as differentiating displacement twice.

$$a = \frac{dv}{dt} = \frac{d^2x}{dt^2}$$

For differentiation with respect to time there is an alternative notation in common usage, using dots over the variables. You write
$v = \dot{x}$
and $a = \dot{v} = \ddot{x}$
The examination may make use of this notation.

EXAMPLE 1

A particle moves in a straight line so that at time t seconds its displacement x metres from an origin O is given by $x = t^4 - 32t$
Find
a where the particle comes instantaneously to rest
b its acceleration at that moment.

a Differentiate to find an expression for the velocity at time t:

$$v = \frac{dx}{dt} = 4t^3 - 32$$

The particle is at rest when $v = 0$.
This gives $4t^3 - 32 = 0$ and so $t = 2$ s
The position of the particle at this instant is

$$x = 2^4 - 32 \times 2 = -48$$

So the particle comes instantaneously to rest at -48 m.

See C1 and C2 for revision of differentiation.

The minus sign indicates that the particle has moved in a negative direction.

b Differentiate again to find an expression for the acceleration:

$$a = \frac{dv}{dt} = 12t^2$$

When $t = 2$, the acceleration a is $48\ \text{m s}^{-2}$

$12 \times 2^2 = 48$

The relationships between x, v and a can be written in terms of integration.

$$v = \frac{dx}{dt} \quad \text{so} \quad x = \int v\,dt$$

$$a = \frac{dv}{dt} \quad \text{so} \quad v = \int a\,dt$$

From **M1** you know that the area under a velocity–time graph represents displacement, and the area under an acceleration–time graph represents change of velocity. In general, you use integration to find the area under a curve.

EXAMPLE 2

A particle travels along a straight wire. It starts from point A with velocity $6\,\text{m s}^{-1}$. Its acceleration is given by $a = -6t\,\text{m s}^{-2}$.

a Find its velocity when $t = 1, 2, 3$ seconds.

b Find its position when $t = 1, 2, 3$ seconds.

c Find its displacement from A during the first 3 seconds of its motion.

d Find how far it travels in the first 3 seconds of its motion.

See **C1** and **C2** for revision of integration.

a To obtain velocity, integrate acceleration:

$$v = \int -6t\,dt = -3t^2 + c$$

When $t = 0$, $v = 6$, which gives $c = 6$ so $v = 6 - 3t^2$

Remember the constant of integration.

Substitute $t = 1$, 2 and 3:

t	1	2	3
v	3	-6	-21

b To obtain displacement, integrate velocity:

$$x = \int 6 - 3t^2\,dt = 6t - t^3 + k$$

When $t = 0$, $x = 0$, which gives $k = 0$ so $x = 6t - t^3$

Substitute $t = 1$, 2 and 3:

t	1	2	3
x	5	4	-9

c The particle is at A when $t = 0$ and at -9 m when $t = 3$. It undergoes a displacement of -9 m.

You could have found the displacement in part **c** using the definite integral

$$\int_0^3 6 - 3t^2\,dt$$

This does **not** give the distance travelled.

d The diagram shows the motion of the particle.

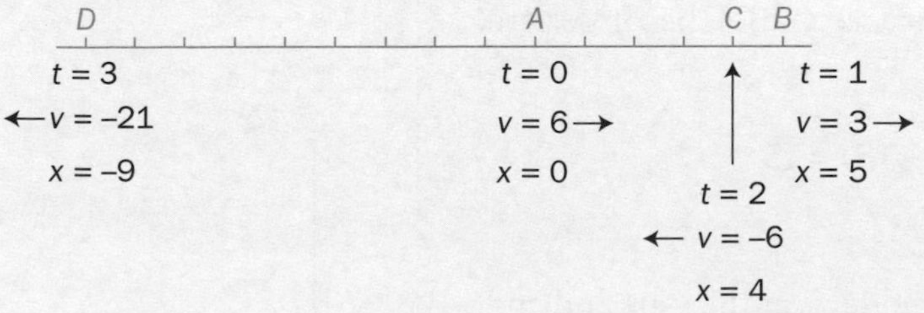

At $t = 1$ the particle passes through B. At that stage, it is moving to the right at $3\,\text{m s}^{-1}$. During the next second it stops momentarily before returning through C and A and finally travelling to D. (After this it continues to travel to the left at increasing speed.)

Example 2 continues on the next page.

EXAMPLE 2 (CONT.)

To find the total distance travelled, you need to know where it comes to rest.

At this point, $v = 0$ so $6 - 3t^2 = 0$

giving $t = \sqrt{2}$ (or $-\sqrt{2}$, which is inappropriate)

When $t = \sqrt{2}$, $x = 6 \times \sqrt{2} - (\sqrt{2})^3 = 4\sqrt{2}$

Total distance is $(9 + 8\sqrt{2})$ m

The particle moves $4\sqrt{2}$ m to the right, then $4\sqrt{2}$ m back to *A* and a further 9 m from *A* to *D*.

$(4\sqrt{2} + 4\sqrt{2} + 9) = 9 + 8\sqrt{2}$

EXAMPLE 3

A car travels between two sets of traffic lights 300 m apart. It starts from rest at the first set of lights, accelerates to a maximum speed before slowing down to stop at the second set. The whole journey takes 30 s. As the velocity is zero when $t = 0$ and when $t = 30$, it is suggested that a possible model is $v = kt(30 - t)$ where $k > 0$.

a Find an expression for the displacement at time t.

b Calculate the value of k.

c Find an expression for the acceleration at time t.

d Calculate the maximum speed reached by the car.

e Criticise the model.

---

a Integrate velocity to obtain displacement:

$$x = \int kt(30 - t)\,dt = k\left(15t^2 - \frac{1}{3}t^3\right) + c$$

$x = 0$ when $t = 0$ which gives $c = 0$

so you have $x = k\left(15t^2 - \frac{1}{3}t^3\right) = \frac{1}{3}kt^2(45 - t)$

b $x = 300$ when $t = 30$ so $300 = \frac{1}{3}k \times 30^2(45 - 30)$

which gives $k = \frac{1}{15}$

c Differentiate velocity to obtain acceleration:

$$v = \frac{1}{15}t(30 - t) \quad \text{so} \quad a = \dot{v} = 2 - \frac{2}{15}t$$

$\dot{v}$ means $\frac{dv}{dt}$

d Maximum speed occurs when acceleration becomes zero.

When $2 - \frac{2}{15}t = 0$, $t = 15$

When $t = 15$, $v = \frac{1}{15} \times 15(30 - 15) = 15$

So the maximum speed is 15 m s^{-1}.

e The main problem with the model is at the start and end. The car is at rest, but then suddenly accelerates at 2 m s^{-2}. You would expect the acceleration of a car to build gradually from zero.

Similarly, you would expect deceleration to decrease to zero as the car comes to rest, but the model predicts an acceleration of -2 m s^{-2} at that point.

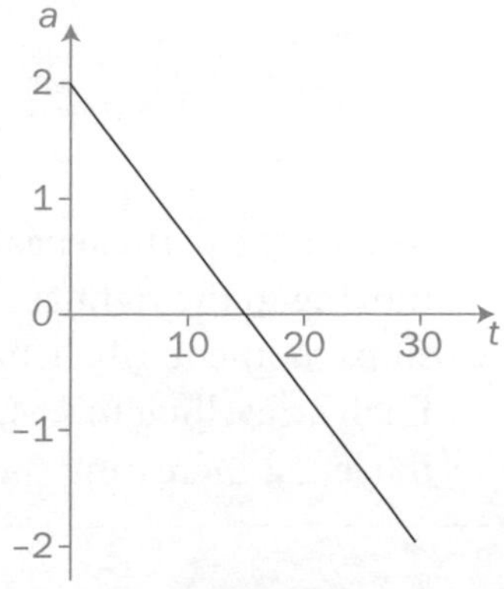

EXAMPLE 4

A particle of mass 3 kg, initially at rest, is acted on by a force $F = (6t + 10)N$ between $t = 0$ s and $t = 10$ s.
Find

a the velocity of the particle at the end of this period

b the distance that it travels.

---

a Call the acceleration a and apply Newton's second law:

$$6t + 10 = 3a$$

giving $a = 2t + \frac{10}{3}$

Newton's second law is $\mathbf{F} = ma$

Integrate acceleration to obtain velocity:

$$v = \int 2t + 3\frac{1}{3}\,dt = t^2 + 3\frac{1}{3}t + c$$

The particle was initially at rest so $v = 0$ when $t = 0$, giving $c = 0$.

Hence $v = t^2 + 3\frac{1}{3}t$

At the end of the period $t = 10$, giving a velocity of $133\frac{1}{3}$ m s^{-1}.

$10^2 + 3\frac{1}{3} \times 10 = 133\frac{1}{3}$

b Integrate velocity to obtain displacement:

$$x = \int t^2 + 3\frac{1}{3}t\,dt = \frac{1}{3}t^3 + 1\frac{2}{3}t^2 + k$$

$x = 0$ when $t = 0$, so $k = 0$

Hence $x = \frac{1}{3}t^3 + 1\frac{2}{3}t^2$

At the end of the period $t = 10$, giving a displacement of 500 m.
So the particle travels a distance of 500 m in the 10 s.

x is the displacement, but the motion was all in the same direction, so it also gives the distance. This is not true in all cases.

M2

Exercise 1.2

1 A particle, moving in a straight line, starts from rest at O. Its acceleration (in m s^{-2}) at time t is given by $a = 30 - 6t$

a Find its velocity and position at time t.

b Find its velocity and position after 5 seconds.

c Find its greatest positive displacement from O.

d Find how long the particle takes to return to O.

2 A particle is moving in a straight line. Its velocity at time t s is given by $v = (6t - 3t^2)$ m s^{-1}

a Find its displacement from time $t = 1$ to time $t = 3$.

b Find the distance it travels from time $t = 1$ to time $t = 3$.

3 A particle, P, is moving along a straight wire. At time t seconds, its displacement x metres from a fixed point, O, on the wire is given by $x = t(t^2 - 16)$
Find

a the time(s) when P is at the fixed point O

b the time(s) when P is not moving

c the displacement of P from O when P is stationary

d the acceleration of P when $t = 5$.

4 The velocity of a particle, travelling along a straight line, is given by $v = 4t + 6$, where the positive direction is to the right. At time $t = 0$, the particle is 8 m to the left of point A.

a Find an expression for the position of the particle at time t.

b Find at what time the particle is at the point A.

c Part **b** gave two values, one of which you disregarded.
Explain what meaning might be attached to the discarded value.

M2

5 A particle is moving along a straight line and its position, measured from the point O, is given by the formula

$$x = t^3 - 2t^2 - t + 2$$

where x is measured in metres and t is measured in seconds. Find

a the times when the particle is at O

b the velocity and acceleration of the particle at the times when it is at O.

6 A particle moving along a straight line through a point O has acceleration given by $a = (2t - 5)\,\text{m s}^{-2}$. When $t = 4$, the particle has velocity $2\,\text{m s}^{-1}$ and its displacement from O is $+8$ m. Find

a an expression for its velocity at time t

b when the particle is at rest

c where the particle is when it is at rest.

7 A bird leaves its nest and flies along a straight line to an adjacent tree, where it collects some food (without landing). It then returns to its nest along the same line. Its position relative to its nest is modelled by the formula

$x = 30t - t^2$ where x is in metres and t is in seconds.

a How long does the journey take?

b How far away is the second tree?

c Criticise the model.

8 A ball is thrown straight up in the air. Its height, at time t s, is h m above the ground, where $h = 4 + 8t - 5t^2$

a Find how long it takes the ball to reach the ground.

b Find an expression for its velocity, v.

c Find its velocity when it reaches the ground.

d Find the maximum height reached by the ball.

e Explain why the expression $\int_0^t v\,dt$ does not represent the distance travelled by the ball.

9 A particle moves in a straight line so that its displacement, x m, at time t s from an origin O is given by $x = t^2 \sin 2t$

a Find the first two positive non-zero times, t_1 and t_2, for which the particle is at O.

b Show that the velocity of the particle at time t_1 is $-\frac{\pi^2}{2}$ m s^{-1}, and find its velocity at time t_2.

c Find the acceleration of the particle at these two times.

10 A particle of mass 5 kg, initially at rest, is acted on for 3 s by a variable force $(40t + 10)$ N. Find

a the final velocity of the particle

b the distance the particle travels while the force is acting.

11 A particle of mass 2 kg is initially travelling at 18 m s^{-1}. It is acted upon by a resistive force of magnitude $2t$ N which brings it to rest. Calculate

a the length of time it takes to stop

b the distance it travels before coming to rest.

12 Example 3 on page 14 explored a simple and rather unsatisfactory model for the motion of a car between two sets of traffic lights.

a Devise an improved model meeting these requirements:

- The car travels 300 m between the lights in 30 s
- Velocity is zero at the start and the end, and is a maximum after 15 s
- Acceleration is zero at the start and the end (and at 15 s for maximum velocity)

b Calculate the maximum velocity using your model.

1.3 Motion in a plane

In one-dimensional motion, you differentiated displacement with respect to time to obtain an expression for velocity. Similarly, you differentiated velocity to obtain acceleration. For a particle moving in a plane the same relationships apply, but the position, velocity and acceleration are expressed as vectors.

To differentiate a vector you need to differentiate each of its components.

For example, the x-component of velocity is the rate of change of displacement in the x-direction.

Displacement	$\mathbf{r} = x\mathbf{i} + y\mathbf{j}$
Velocity	$\mathbf{v} = \dot{\mathbf{r}} = \dot{x}\mathbf{i} + \dot{y}\mathbf{j}$
Acceleration	$\mathbf{a} = \ddot{\mathbf{r}} = \ddot{x}\mathbf{i} + \ddot{y}\mathbf{j}$

$\dot{\mathbf{r}}$ means $\frac{d\mathbf{r}}{dt}$

$\ddot{\mathbf{r}}$ means $\frac{d^2\mathbf{r}}{dt^2}$ etc.

The components must be functions of time.
To emphasise this you sometimes write $\mathbf{r} = f(t)\mathbf{i} + g(t)\mathbf{j}$

Position vector	$\mathbf{r} = f(t)\mathbf{i} + g(t)\mathbf{j}$
Velocity	$\mathbf{v} = \dot{\mathbf{r}} = f'(t)\mathbf{i} + g'(t)\mathbf{j}$
Acceleration	$\mathbf{a} = \ddot{\mathbf{r}} = f''(t)\mathbf{i} + g''(t)\mathbf{j}$

$f'(t)$ means $\frac{d}{dt}f(t)$,

$f''(t)$ means $\frac{d^2}{dt^2}f(t)$, etc.

EXAMPLE 1

The displacement, in m, of a particle at time t s is given by

$\mathbf{r} = 4t^2\mathbf{i} + (3t - 5t^3)\mathbf{j}$

Find

a its speed when $t = 1$

b the direction in which it is accelerating at that time.

a Differentiate $\mathbf{r}$ to obtain the velocity:

$$\mathbf{v} = \dot{\mathbf{r}} = 8t\mathbf{i} + (3 - 15t^2)\mathbf{j}$$

When $t = 1$, you have $\mathbf{v} = (8\mathbf{i} - 12\mathbf{j})\ \text{m s}^{-1}$

Speed is the magnitude of velocity, so

$$\text{speed} = \sqrt{8^2 + (-12)^2}\ \text{m s}^{-1} = 4\sqrt{13}\ \text{m s}^{-1}$$

b Differentiate a second time to obtain acceleration:

$$\mathbf{a} = \ddot{\mathbf{r}} = 8\mathbf{i} - 30t\mathbf{j}$$

When $t = 1$, you have $\mathbf{a} = (8\mathbf{i} - 30\mathbf{j})\ \text{m s}^{-2}$

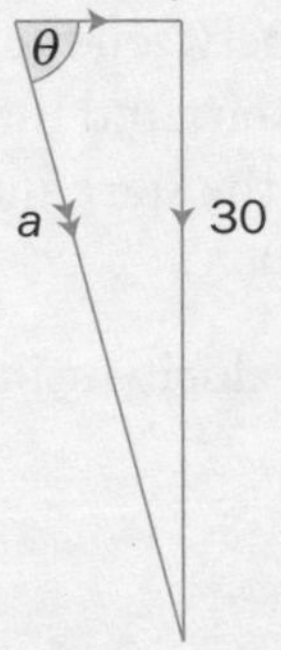

A diagram helps to visualise the situation.

In the diagram $\tan\theta = \frac{30}{8}$ so $\theta = 75.1°$

So the direction of acceleration is at $-75.1°$ to the $\mathbf{i}$-direction.

As with one-dimensional motion, for motion in a plane you also have

$$\mathbf{v} = \int \mathbf{a}\,\mathrm{d}t$$

$$\mathbf{r} = \int \mathbf{v}\,\mathrm{d}t$$

To integrate a vector you must integrate each of its components.

EXAMPLE 2

The acceleration of a particle at time t is $\mathbf{a} = 4\mathbf{i} + 6t\mathbf{j}$
Find an expression for its velocity.

Integrate acceleration to obtain velocity:

$$\mathbf{v} = \int 4\mathbf{i} + 6t\mathbf{j}\,\mathrm{d}t = (4t + c_1)\mathbf{i} + (3t^2 + c_2)\mathbf{j}$$

Each component has a constant of integration.

Gather together the constant parts:

$$\mathbf{v} = 4t\mathbf{i} + 3t^2\mathbf{j} + (c_1\mathbf{i} + c_2\mathbf{j})$$

$$\mathbf{v} = 4t\mathbf{i} + 3t^2\mathbf{j} + \mathbf{c}$$

where $\mathbf{c}$ is a constant vector.

You can find the value of this constant vector if you know corresponding values of $\mathbf{v}$ and t.

M2

EXAMPLE 3

The velocity of a particle at time t is given by

$$\mathbf{v} = 6t^2\mathbf{i} + (3t^2 - 8t)\mathbf{j}$$

Find its position vector at time t, given that $\mathbf{r} = 5\mathbf{i} - 6\mathbf{j}$ when $t = 0$.

Integrate velocity to obtain displacement:

$$\mathbf{r} = \int 6t^2\mathbf{i} + (3t^2 - 8t)\mathbf{j}\,\mathrm{d}t$$

$$= 2t^3\mathbf{i} + (t^3 - 4t^2)\mathbf{j} + \mathbf{c}$$

To find $\mathbf{c}$, use the condition $\mathbf{r} = 5\mathbf{i} - 6\mathbf{j}$ when $t = 0$:

$$5\mathbf{i} - 6\mathbf{j} = 0\mathbf{i} + 0\mathbf{j} + \mathbf{c}$$

and so $\mathbf{c} = 5\mathbf{i} - 6\mathbf{j}$

Substitute for $\mathbf{c}$ in the general expression:

$$\mathbf{r} = 2t^3\mathbf{i} + (t^3 - 4t^2)\mathbf{j} + 5\mathbf{i} - 6\mathbf{j}$$

or $\mathbf{r} = (2t^3 + 5)\mathbf{i} + (t^3 - 4t^2 - 6)\mathbf{j}$

EXAMPLE 4

A particle moves in a plane with an acceleration of $2t\mathbf{j}\ \mathrm{m\,s^{-2}}$.
At time $t = 0$ the particle is at the point $(\mathbf{i} + 4\mathbf{j})$ m and has velocity $(3\mathbf{i} - 4\mathbf{j})\ \mathrm{m\,s^{-1}}$.
Find expressions for

a the velocity of the particle

b the position of the particle at time t.

---

a Integrate acceleration to obtain velocity:

$$\mathbf{v} = \int 2t\mathbf{j}\,\mathrm{d}t = t^2\mathbf{j} + \mathbf{c}$$

To find $\mathbf{c}$, use the condition $\mathbf{v} = 3\mathbf{i} - 4\mathbf{j}$ when $t = 0$:

$$3\mathbf{i} - 4\mathbf{j} = 0\mathbf{j} + \mathbf{c} \quad \text{and so} \quad \mathbf{c} = 3\mathbf{i} - 4\mathbf{j}$$

The velocity is therefore $\mathbf{v} = t^2\mathbf{j} + 3\mathbf{i} - 4\mathbf{j} = 3\mathbf{i} + (t^2 - 4)\mathbf{j}$

b Integrate velocity to obtain displacement:

$$\mathbf{r} = \int 3\mathbf{i} + (t^2 - 4)\mathbf{j}\,\mathrm{d}t = 3t\mathbf{i} + \left(\frac{1}{3}t^3 - 4t\right)\mathbf{j} + \mathbf{c}_1$$

To find $\mathbf{c}_1$, use the condition $\mathbf{r} = \mathbf{i} + 4\mathbf{j}$ when $t = 0$:

$$\mathbf{i} + 4\mathbf{j} = 0\mathbf{i} + 0\mathbf{j} + \mathbf{c}_1 \quad \text{and so} \quad \mathbf{c}_1 = \mathbf{i} + 4\mathbf{j}$$

The position is therefore

$$\mathbf{r} = 3t\mathbf{i} + \left(\frac{1}{3}t^3 - 4t\right)\mathbf{j} + \mathbf{i} + 4\mathbf{j} = (3t + 1)\mathbf{i} + \left(\frac{1}{3}t^3 - 4t + 4\right)\mathbf{j}$$

EXAMPLE 5

A particle of mass 4 kg is acted on by a force $(8\mathbf{i} + 12t\mathbf{j})$ N.
Initially the particle has velocity $(3\mathbf{i} - 5\mathbf{j})\ \mathrm{m\,s^{-1}}$.
Find its velocity after 4 seconds.

The equation of motion of the particle is

$$4\mathbf{a} = 8\mathbf{i} + 12t\mathbf{j} \quad \text{which gives} \quad \mathbf{a} = 2\mathbf{i} + 3t\mathbf{j}$$

Using Newton's second law: $\mathbf{F} = m\mathbf{a}$

Integrate acceleration to obtain velocity:

$$\mathbf{v} = \int 2\mathbf{i} + 3t\mathbf{j}\,\mathrm{d}t = 2t\mathbf{i} + \frac{3}{2}t^2\mathbf{j} + \mathbf{c}$$

From the condition $\mathbf{v} = 3\mathbf{i} - 5\mathbf{j}$ when $t = 0$
you obtain $\mathbf{c} = 3\mathbf{i} - 5\mathbf{j}$

So
$$\mathbf{v} = 2t\mathbf{i} + \frac{3}{2}t^2\mathbf{j} + 3\mathbf{i} - 5\mathbf{j}$$
$$= (2t + 3)\mathbf{i} + \left(\frac{3}{2}t^2 - 5\right)\mathbf{j}$$

Substitute $t = 4$: $\mathbf{v} = 11\mathbf{i} + 19\mathbf{j}$

Exercise 1.3

1 For each of the following position vectors, find expressions for the velocity ($\dot{\mathbf{r}}$) and acceleration ($\ddot{\mathbf{r}}$) of the particle at time t.

a $\mathbf{r} = 4t\mathbf{i} + (8 + 2t^2)\mathbf{j}$

b $\mathbf{r} = (t^2 - 4t)\mathbf{i} + (t^3 - 2t^2)\mathbf{j}$

c $\mathbf{r} = 4\sqrt{t}\mathbf{i} + \frac{1}{4t^2}\mathbf{j}$

d $\mathbf{r} = (t^2 + 1)^3\mathbf{i} + \frac{1}{2t + 1}\mathbf{j}$

e $\mathbf{r} = 2\cos t\,\mathbf{i} + 2\sin t\,\mathbf{j}$

f $\mathbf{r} = \mathrm{e}^t\mathbf{i} + \ln(t + 1)\mathbf{j}$

2 For each of the following velocity vectors, find expressions for the displacement and acceleration at time t, with arbitrary constants where necessary.

a $\mathbf{v} = t^3\mathbf{i} + 3t^2\mathbf{j}$

b $\mathbf{v} = 15\mathbf{i} + (20 - 10t)\mathbf{j}$

c $\mathbf{v} = (t - t^2)\mathbf{i} + (3t - 5)\mathbf{j}$

d $\mathbf{v} = 2t(1 - 3t)\mathbf{i} + t^2(3 - 4t)\mathbf{j}$

3 For each of the following acceleration vectors, find expressions for the velocity and displacement at time t, with arbitrary constants where necessary.

a $\mathbf{a} = 6t\mathbf{i} + (2 - 4t)\mathbf{j}$

b $\mathbf{a} = 2\mathbf{i} + 4t\mathbf{j}$

c $\mathbf{a} = \frac{4}{t^3}\mathbf{i} + 3\sqrt{t}\mathbf{j}$

d $\mathbf{a} = -12\cos 2t\mathbf{i} - 12\sin 2t\mathbf{j}$

4 For the following velocity vectors, find the position vector $\mathbf{r}$ at time t consistent with the given initial condition.

a $\mathbf{v} = 4t\mathbf{i} + 8t^3\mathbf{j}$, given that $\mathbf{r} = 2\mathbf{i} - \mathbf{j}$ when $t = 0$

b $\mathbf{v} = 6\sin t\mathbf{i} + 2\mathbf{j}$, given that $\mathbf{r} = 3\mathbf{i} + 2\mathbf{j}$ when $t = 0$

5 For the following acceleration vectors, find the velocity vector $\mathbf{v}$ and position vector $\mathbf{r}$ at time t consistent with the given initial conditions.

a $\mathbf{a} = (3 - 2t)\mathbf{i} + (2t - 6t^3)\mathbf{j}$, given that $\mathbf{v} = 3\mathbf{i}$ and $\mathbf{r} = \mathbf{i} - 2\mathbf{j}$ when $t = 0$

b $\mathbf{a} = 4\cos 2t\mathbf{i} + 8\sin 2t\mathbf{j}$, given that $\mathbf{v} = 2\mathbf{i} + \mathbf{j}$ and $\mathbf{r} = 4\mathbf{j}$ when $t = 0$

6 A particle moves on a plane such that its position at time t is given by

$$\mathbf{r} = (3t - 2)\mathbf{i} + (4t - 2t^2)\mathbf{j}$$

a Find expressions for the velocity and acceleration of the particle at time t.

b Find the initial speed of the particle.

c At what time(s) is the particle moving parallel to the x-axis?

d Is the particle ever stationary? Give a reason for your answer.

7 The position of a particle at time t s is $\mathbf{r} = 2t^3\mathbf{i} + t^2\mathbf{j}$ m
Find the speed of the particle when $t = 1$.

M2

8 A particle moves in a plane, starting from the point whose position vector is $2\mathbf{i} + 3\mathbf{j}$. Its velocity $\mathbf{v}$ m s^{-1} at time t s is given by

$\mathbf{v} = (4t - 2)\mathbf{i} + 3t^2\mathbf{j}$

a Find an expression for the position of the particle at time t.

b Find the average velocity of the particle over the interval $t = 0$ to $t = 3$.

✓ 9 The acceleration, $\mathbf{a}$ m s^{-2}, of a particle P at time t seconds is given by $\mathbf{a} = 6t\mathbf{i} + 2\mathbf{j}$

When $t = 0$ the velocity of P is $(3\mathbf{i} + 4\mathbf{j})$ m s^{-1}.

a Show that, when $t = 2$, P is moving at 17 m s^{-1}.

b When $t = 2$, P is moving at an angle α to the $\mathbf{i}$-direction. Find the value of α.

10 A particle moves in a plane. Its acceleration is $\mathbf{a} = -2\mathbf{j}$

At time $t = 0$ the particle is at the point with position vector $2\mathbf{i} - 3\mathbf{j}$ and has velocity $2\mathbf{i} + 4\mathbf{j}$.

a Find expressions for its velocity and position at time t.

b At what time(s) is the particle moving in the $\mathbf{i}$-direction?

c At what time(s) does the particle cross the x-axis?

M2

✓ 11 A particle of mass 5 kg is acted upon by a force $(20t\mathbf{i} - 15\mathbf{j})$ N. Initially the particle has position vector $(2\mathbf{i} + 3\mathbf{j})$ m and travels with velocity $(-2\mathbf{i} + 12\mathbf{j})$ m s^{-1}. Find expressions for the velocity and position of the particle at time t, and hence find its velocity and position when $t = 6$.

✓ 12 A particle of mass 3 kg is acted upon by a variable force $(6\mathbf{i} + 6(t - 1)\mathbf{j})$ N. Initially the particle is at the origin and is travelling with velocity $-3\mathbf{j}$ m s^{-1}. Show that at one subsequent time it is travelling in the $\mathbf{i}$-direction, and find its speed and position at that moment.

✓ 13 Take east and north to be the $\mathbf{i}$- and $\mathbf{j}$-directions.

A particle of mass 2 kg is acted upon by two forces $\mathbf{P}_1$ N and $\mathbf{P}_2$ N, where $\mathbf{P}_1 = 5\mathbf{i} - 2t\mathbf{j}$ and $\mathbf{P}_2 = (2t - 9)\mathbf{i} + 6t\mathbf{j}$

Initially the particle is at rest.

a Find an expression for the velocity of the particle at time t.

b At what time is the particle travelling due north?

c Show that the particle never travels in a north-easterly direction, and find the time at which it is travelling in a north-westerly direction.

14 Two particles have velocities $\mathbf{v}_1 = ((2 + 2t)\mathbf{i} + (t + 2)\mathbf{j})\text{ m s}^{-1}$ and $\mathbf{v}_2 = (3\mathbf{i} - 4\mathbf{j})\text{ m s}^{-1}$. Initially they have position vectors $\mathbf{r}_1 = (2\mathbf{i} + \mathbf{j})\text{ m}$ and $\mathbf{r}_2 = (-\mathbf{i} + \mathbf{j})\text{ m}$

a Find the time at which the particles are moving perpendicular to each other.

b Find the distance between the particles at this time.

15 A particle of mass 3 kg moves under the action of a force of $(6\mathbf{i} - 36t^2\mathbf{j})$ N. The particle is initially at rest at the point with position vector $(\mathbf{i} + \mathbf{j})$ m.

a Find the velocity and position of the particle at time t.

b If the position of the particle at time t is $(x\mathbf{i} + y\mathbf{j})$ m, find an equation connecting x and y, and deduce that the path of the particle is a parabola.

16 Two particles, A and B, leave the origin at the same time and move on a plane so that at time t their position vectors are

$$\mathbf{r}_A = (8t - 2t^2)\mathbf{i} + (4t + t^2)\mathbf{j} \quad \text{and} \quad \mathbf{r}_B = (4t + t^2)\mathbf{i} + \left(6t + 1\tfrac{1}{2}t^2\right)\mathbf{j}$$

Find at what times the particles are moving in the same or opposite directions.

17 The position vector of a particle is given by

$$\mathbf{r} = (2 + 3t + 8t^2)\mathbf{i} + (6 + t + 12t^2)\mathbf{j}$$

a Find an expression for the velocity of the particle at time t.

b Find an expression for the acceleration of the particle at time t.

An observer placed at the origin sees the particle in the direction given by $\tan\theta = \frac{y}{x}$, where $\mathbf{r} = x\mathbf{i} + y\mathbf{j}$ is the position vector of the particle. The direction in which the particle is moving is given by $\tan\phi = \frac{b}{a}$, where $\mathbf{v} = a\mathbf{i} + b\mathbf{j}$ is the velocity of the particle. The particle will be moving directly away from or directly towards the observer when $\tan\theta = \tan\phi$

c Find the time(s) when the particle is moving directly towards or directly away from the observer.

d Find the position of the particle at the time(s) found in part **c** and identify whether the particle is moving towards or away from the observer.

1 A golfer hits a ball from a point O on a horizontal surface. The initial velocity of the ball is $28\,\text{m s}^{-1}$ at an angle of $50°$ to the horizontal.

a Calculate the maximum height reached by the ball.

b Calculate the range of the ball.

c How fast was the ball travelling at the highest point of its trajectory?

d What assumptions have you made in modelling this situation?

2 Ayesha kicks a ball from a point O on horizontal ground towards a wall of height 2.5 m which is 12 m from O. The initial velocity of the ball is $(10\mathbf{i} + 8\mathbf{j})\,\text{m s}^{-1}$.

a How long does the ball take to reach the wall?

b Investigate whether the ball strikes or passes over the wall.

c Find the speed and direction of travel of the ball at the moment that it reaches the wall.

3 A stone is thrown from the top of a vertical cliff, d m above a horizontal beach. The initial velocity of the stone is $15\,\text{m s}^{-1}$ at an angle θ above the horizontal, where $\sin\theta = \frac{3}{5}$. The stone lands on the beach at a distance of 48 m from the base of the cliff.

a Calculate the time of flight of the stone.

b Calculate the value of d.

c Find the direction in which the stone is travelling at the instant that it strikes the beach.

4 Jermaine kicks a ball from a point O on the floor of a sports hall towards the wall which is 30 m from O. The ceiling is at a height of 8 m. The initial velocity of the ball is $V\,\text{m s}^{-1}$ at an angle of θ to the horizontal.

a Show that the ball will just touch the ceiling if $V\sin\theta = 4\sqrt{g}$

b Find a second relationship between V and θ if the ball lands exactly at the base of the wall.

Jermaine manages to kick the ball so that it just brushes the ceiling and lands exactly at the base of the wall.

c Show that $\tan\theta = \frac{16}{15}$ and find the value of V.

5 Nancy is in a small boat in a bank of fog which is 15 m deep. Her outboard motor has failed so she fires a distress flare. The initial velocity of the flare is $28\,\text{m s}^{-1}$ at an angle of θ to the horizontal. Assume that the flare is a particle and there is no air resistance.

a Show that the greatest height reached by the flare is $40\sin^2\theta\,\text{m}$.

b For what values of θ will the flare be visible above the fog bank?

c Nancy fires the flare at 60° to the horizontal. For how long is the flare visible above the fog bank?

6 Jim is standing at the top of a tower, which stands on horizontal ground. He throws a ball to Samantha, who is standing 18 m from the base of the tower. He throws the ball at $12\,\text{m s}^{-1}$ from a point 20 m above the ground. The initial direction of the ball is 20° below the horizontal.

a How long does the ball take to reach Samantha?

b At what height above the ground does she catch the ball?

c Find the speed and direction of the ball at the moment that she catches it.

7 A particle starts from rest at a point O and travels along a straight line. Its displacement from O at time t s is x m.
Its displacement is modelled by $x = \frac{3t^2}{10} + \frac{3t^3}{100} - \frac{t^4}{500}$

a Find the displacement of the particle from O after 10 s.

b Find an expression for the velocity of the particle at time t and hence show that after 5 s it is travelling at $4.25\,\text{m s}^{-1}$.

c Find an expression for the acceleration of the particle at time t and hence find its acceleration when $t = 10$.

d Calculate the maximum velocity of the particle.

8 A particle moves in a straight line so that its displacement at time t s from a point O on the line is x m. Its velocity, $v\,\text{m s}^{-1}$, at time t is modelled by $v = 12t - 3t^2$

a Find the maximum velocity of the particle.

b The particle starts from O. Find an expression for x at time t. Hence find the displacement undergone by the particle between $t = 2$ and $t = 5$.

c Find where and when the particle is instantaneously at rest.

d Find the distance travelled by the particle between $t = 2$ and $t = 5$.

9 A particle moves along a straight line. At time t the displacement of the particle from its initial position is x where

$$x = 4t^2 + 4e^{-2t} + 7$$

a Find the velocity of the particle at time t.

b Find the acceleration of the particle at time t.

c Describe what happens to the acceleration of the particle as t increases.

10 A particle moves in a straight line so that at time t s its acceleration is $(7 - 4t)$ m s^{-2}. When $t = 0$ the particle is at a point O and has velocity 4 m s^{-1}.

a Find an expression for the velocity v at time t.

b Show that the maximum velocity of the particle is $10\frac{1}{8}$ m s^{-1}.

c Find an expression for the displacement, x m, of the particle from O at time t.

d Find the greatest positive displacement of the particle from O during the motion.

M2

11 A train of mass 5 tonnes is travelling at 30 m s^{-1} when the brakes are progressively applied so that after t s there is a retarding force of magnitude $3000t$ N.

a Write down an expression for the acceleration, a m s^{-2}, of the train t s after the brakes were applied.

b Find an expression for the velocity, v m s^{-1}, of the train at time t.

c Find how long the train takes to come to rest.

d How far does the train travel during this time?

12 A particle P moves so that at time t s its position vector, $\mathbf{r}$ m, is

$$\mathbf{r} = (t^3 - 3t)\mathbf{i} + (t^2 - 4t)\mathbf{j}$$

a Find an expression for the velocity, $\mathbf{v}$ m s^{-1}, of the particle at time t.

b Show that the particle is never stationary.

c For what value of t is the particle travelling parallel to the x-axis? Show that it is $2\sqrt{5}$ m from the origin at that instant.

13 A force, $\mathbf{F} = (2000t\mathbf{i} - 4000\mathbf{j})$ N acts on a particle of mass 500 kg at time t seconds.

a Find the acceleration, $\mathbf{a}\,\mathrm{m\,s^{-2}}$, of the particle.

b At time $t = 0$, the velocity of the particle is $10\mathbf{i}\,\mathrm{m\,s^{-1}}$. Show that the velocity, $\mathbf{v}\,\mathrm{m\,s^{-1}}$, of the particle at time t is

$$\mathbf{v} = (10 + 2t^2)\mathbf{i} - 8t\mathbf{j}$$

c Initially the particle is at the origin. Find the position vector, $\mathbf{r}$ m, of the particle at time t.

d Find the distance of the particle from the origin when $t = 3$.

14 A particle P moves in a plane so that its position vector at time t s relative to an origin O is

$$\mathbf{r} = (\cos 2t\mathbf{i} + \sin 2t\mathbf{j})\ \mathrm{m}$$

a Calculate the distance of P from O at time ts and hence describe the path of the particle.

b Find an expression for $\mathbf{v}$, the velocity of the particle at time t. Hence show that the particle is moving at constant speed, which you should state.

15 A particle P of mass 2 kg moves in a plane under the action of a force $\mathbf{F}$ N so that its position vector at time ts relative to an origin O is

$$\mathbf{r} = ((t - 2t^2 + 3)\mathbf{i} + (t + 1)^2\mathbf{j})\ \mathrm{m}$$

a Find the displacement undergone by P in the first second of its motion.

b Find the speed and direction of P when $t = 1$.

c Show that $\mathbf{F}$ is constant, and find its magnitude and direction.

16 Particles P and Q, each of mass 0.5 kg, move in the same horizontal plane, with east and north as the $\mathbf{i}$- and $\mathbf{j}$-directions. Initially P has velocity $(2\mathbf{i} - 5\mathbf{j})\,\mathrm{m\,s^{-1}}$, while Q moves at $2\,\mathrm{m\,s^{-1}}$ in a northerly direction. Each particle is acted on by a variable force of magnitude tN at time t s. The force on P acts towards the north-east, while that on Q is towards the south-east.

a Find the value of t for which the two particles are moving with the same speed.

b Find the value of t for which the two particles are moving in the same direction.

1

Exit

Summary

Refer to

- For a projectile
  - horizontal motion is at constant speed, so velocity is $v_x = u\cos\alpha$ and displacement is $x = ut\cos\alpha$
  - vertical motion has acceleration $-g$, so velocity is $v_y = u\sin\alpha - gt$ and displacement is $y = ut\sin\alpha - \frac{1}{2}gt^2$ 1.1
- For a particle moving along a straight line, displacement x, velocity v and acceleration a are functions of time t or constants.
  - Velocity is rate of change of displacement, so
    $v = \frac{dx}{dt}$, also written as $v = \dot{x}$
    Acceleration is rate of change of velocity, so
    $a = \frac{dv}{dt} = \frac{d^2x}{dt^2}$, also written as $a = \dot{v} = \ddot{x}$
  - Reversing the process, $x = \int v\,dt$ and $v = \int a\,dt$
- To differentiate/integrate a vector you differentiate/integrate its components separately. 1.2
- For a particle moving in a plane, the position vector **r**, velocity **v** and acceleration **a** are in general functions of time t.
  - If $\mathbf{r} = f(t)\mathbf{i} + g(t)\mathbf{j}$ then
    $$\mathbf{v} = \frac{d\mathbf{r}}{dt} = \dot{\mathbf{r}} = f'(t)\mathbf{i} + g'(t)\mathbf{j}$$
    $$\mathbf{a} = \frac{d\mathbf{v}}{dt} = \frac{d^2\mathbf{r}}{dt^2} = \ddot{\mathbf{r}} = f''(t)\mathbf{i} + g''(t)\mathbf{j}$$
  - Reversing the process, $\mathbf{r} = \int \mathbf{v}\,dt$ and $\mathbf{v} = \int \mathbf{a}\,dt$ 1.3

Links

In most real-life situations involving projectiles the motion is affected by air resistance and the size of the object. However, the model shown in this chapter provides a basic understanding of the motion of a projectile.

Film stunts often involve vehicles flying through the air. Stunt advisers can model this motion on a projectile and can use this to predict the outcome of the stunt.

M2

2

Centre of mass

This chapter will show you how to
- understand the concept of centre of mass
- find the centre of mass of a system of particles in one or two dimensions
- find the centre of mass of a uniform plane figure by symmetry
- find the centre of mass of a composite lamina
- find the position of a lamina when it is suspended from a single point, or placed on an inclined plane, in equilibrium.

Before you start

You should know how to:

1 Find the total moment of a system of parallel forces.

Check in:

1 A light rod AB of length 4 m rests in a horizontal position on supports at A and B. Particles of mass 2 kg, 4 kg and 6 kg are attached to the rod at 1 m, 2 m and 3 m from A respectively.
Find the total moment of their weights about A and hence find the reaction in the support at B.

2.1 Centre of mass

Every object behaves as if its mass were concentrated at a single point – its centre of mass.

If the object is in a uniform gravitational field, its weight acts through the centre of mass. This determines how it will hang if you suspend it, and where its 'point of balance' is.

In the simplest case, you find the centre of mass of a system of particles placed at various points along a straight line.

Try suspending an irregular piece of card. The same point G will be below the hook whichever corner you hang it from, and the card will balance on this point, as shown.

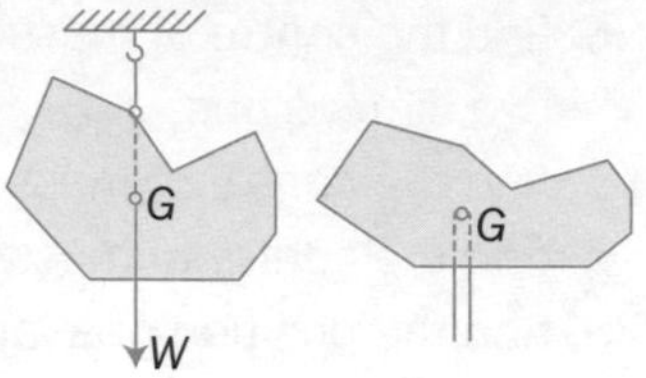

Strictly speaking G is the centre of gravity, but except in extreme contrived situations the two points coincide. The centre of mass is important even in conditions of weightlessness.

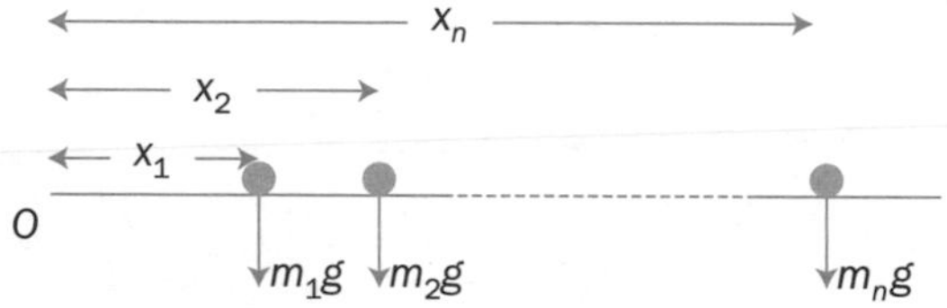

Suppose there are masses $m_1, m_2, \ldots, m_n$ placed along a line with displacements $x_1, x_2, \ldots, x_n$ from an origin O.
The total moment of the system about O is

$$m_1gx_1 + m_2gx_2 + \cdots + m_ngx_n$$

The system is equivalent to a mass $M = m_1 + m_2 + \cdots + m_n$ at the centre of mass G, with displacement $\bar{x}$ from O.

Some of the x-values may be negative – they are only shown positive here to make the illustration simpler.

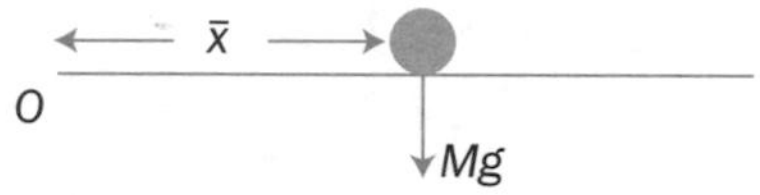

The moment of this system is $Mg\bar{x}$.
The two moments should be the same, so

$$Mg\bar{x} = m_1gx_1 + m_2gx_2 + \cdots + m_ngx_n$$

giving $$\bar{x} = \frac{m_1x_1 + m_2x_2 + \cdots + m_nx_n}{m_1 + m_2 + \cdots + m_n}$$

Cancelling by g and replacing M with $m_1 + m_2 + \cdots + m_n$.

You can write this result using sigma notation.

For a set of masses $m_1, m_2, \ldots, m_n$ placed on a line with displacements $x_1, x_2, \ldots, x_n$ from an origin O, the centre of mass G has displacement $\bar{x}$, where

$$\bar{x} = \frac{m_1x_1 + m_2x_2 + \cdots + m_nx_n}{m_1 + m_2 + \cdots + m_n} = \frac{\sum_{i=1}^{n} m_ix_i}{\sum_{i=1}^{n} m_i}$$

EXAMPLE 1

A light rod AB, of length 4 m, has particles of mass 4 kg, 10 kg and 6 kg attached to it at points 1.5 m, 2 m and 3.5 m from A. Find the position of the centre of mass.

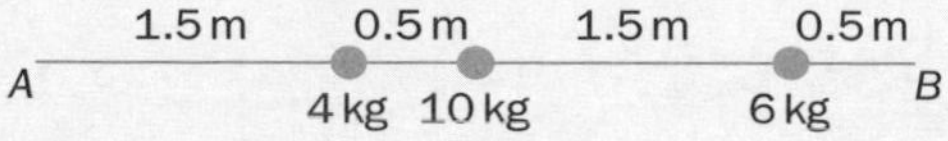

Sketch a diagram.

$$\bar{x} = \frac{\sum_{i=1}^{n} m_i x_i}{\sum_{i=1}^{n} m_i} = \frac{4\times 1.5 + 10\times 2 + 6\times 3.5}{4+10+6} = \frac{47}{20} = 2.35$$

Taking A as the origin.

So the centre of mass is 2.35 m from A.

You can model a beam or strut of uniform density as a **uniform rod** – a one-dimensional figure – if its thickness is negligible compared to its length. By symmetry, you have

The centre of mass, G, of a uniform rod AB is at the midpoint of AB.

M2

EXAMPLE 2

A uniform beam, AB, has length 3 m and mass 30 kg. Particles of mass 10 kg and 40 kg are attached to the beam at C and D, where $AC = 1$ m and $AD = 2.5$ m. Find the position of the centre of mass of the system.

The beam is uniform, so its centre of mass is 1.5 m from A, as shown.

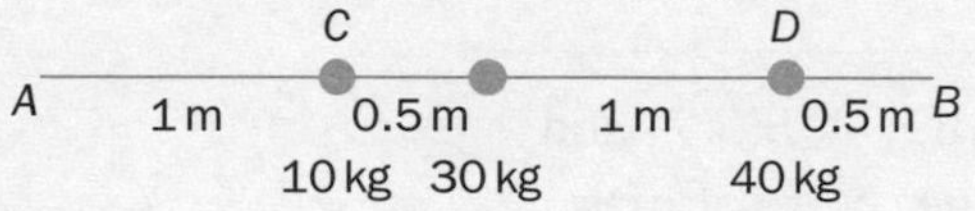

$$\bar{x} = \frac{\sum_{i=1}^{n} m_i x_i}{\sum_{i=1}^{n} m_i} = \frac{10\times 1 + 30\times 1.5 + 40\times 2.5}{10+30+40} = \frac{155}{80} = 1.94$$

Taking A as the origin.

So the centre of mass of the system is 1.94 m from A.

More generally the particles could be placed on a plane.

Particles $m_1, m_2, \ldots, m_n$ are placed on a plane at the points $(x_1, y_1), (x_2, y_2), \ldots, (x_n, y_n)$. The centre of mass is $G(\bar{x}, \bar{y})$.

If the system were in a uniform gravitational field perpendicular to the plane, the resultant weight would act through G.

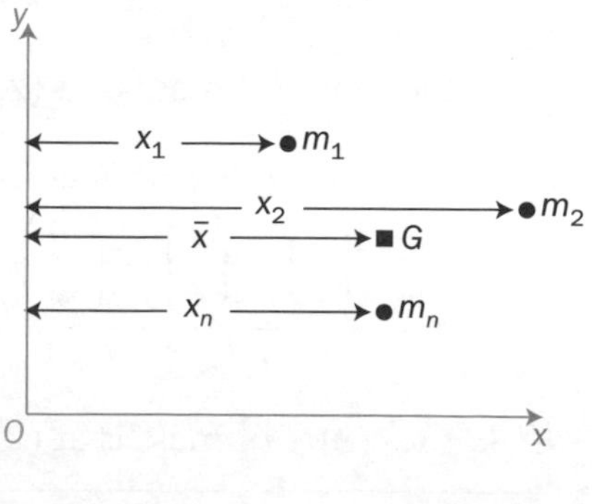

The moment of the resultant weight about the y-axis is the same as the sum of the individual moments of the particles. This gives
$(m_1 + m_2 + \cdots + m_n)g\bar{x} = m_1gx_1 + m_2gx_2 + \cdots + m_ngx_n$

and hence $$\bar{x} = \frac{m_1x_1 + m_2x_2 + \cdots + m_nx_n}{m_1 + m_2 + \cdots + m_n} = \frac{\sum_{i=1}^{n} m_ix_i}{\sum_{i=1}^{n} m_i}$$

Similarly, taking moments about the x-axis, you obtain

$$\bar{y} = \frac{m_1y_1 + m_2y_2 + \cdots + m_ny_n}{m_1 + m_2 + \cdots + m_n} = \frac{\sum_{i=1}^{n} m_iy_i}{\sum_{i=1}^{n} m_i}$$

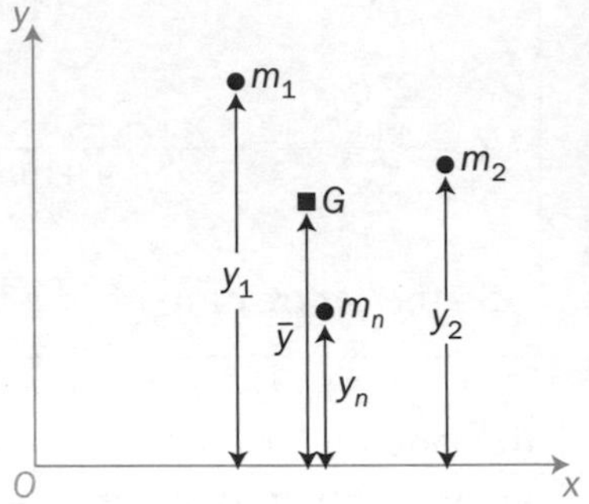

You can express the result in vector notation.

A system of particles of mass $m_1, m_2, \ldots, m_n$ placed at points with position vectors $\mathbf{r}_1 = \begin{pmatrix} x_1 \\ y_1 \end{pmatrix}, \mathbf{r}_2 = \begin{pmatrix} x_2 \\ y_2 \end{pmatrix}, \ldots, \mathbf{r}_n = \begin{pmatrix} x_n \\ y_n \end{pmatrix}$ has centre of mass G with position vector $\bar{\mathbf{r}} = \begin{pmatrix} \bar{x} \\ \bar{y} \end{pmatrix}$, where

$$\begin{pmatrix} \bar{x} \\ \bar{y} \end{pmatrix} = \frac{\sum_{i=1}^{n} m_i \begin{pmatrix} x_i \\ y_i \end{pmatrix}}{\sum_{i=1}^{n} m_i} \quad \text{or} \quad \bar{\mathbf{r}} = \frac{\sum_{i=1}^{n} m_i\mathbf{r}_i}{\sum_{i=1}^{n} m_i}$$

where $\begin{pmatrix} x_1 \\ y_1 \end{pmatrix} = x_1\mathbf{i} + y_1\mathbf{j}$, etc

M2

EXAMPLE 3

Masses of 2 kg, 3 kg and 5 kg are placed at $A(3,1)$, $B(5,7)$ and $C(1,-4)$ respectively. Find the position of the centre of mass.

Take moments about the y-axis:

$$\bar{x} = \frac{2\times3 + 3\times5 + 5\times1}{2+3+5} = 2.6$$

Take moments about the x-axis:

$$\bar{y} = \frac{2\times1 + 3\times7 + 5\times(-4)}{2+3+5} = 0.3$$

So the centre of mass is at (2.6, 0.3).

You could have shown this in vector form:

$$\bar{\mathbf{r}} = \frac{2\begin{pmatrix} 3 \\ 1 \end{pmatrix} + 3\begin{pmatrix} 5 \\ 7 \end{pmatrix} + 5\begin{pmatrix} 1 \\ -4 \end{pmatrix}}{2+3+5} = \begin{pmatrix} 2.6 \\ 0.3 \end{pmatrix}$$

So the centre of mass is at (2.6, 0.3).

EXAMPLE 4

Masses of 2 kg, 4 kg, 5 kg and 3 kg are placed respectively at the vertices A, B, C and D of a light rectangular framework $ABCD$, where $AB = 3$ m and $BC = 2$ m. Further masses of 1 kg and 5 kg are placed at E and F, the midpoints of BC and CD respectively. Find the centre of mass of the system.

The framework in Example 4 is light. (It has negligiblse mass).

Take AB and AD to be the x- and y-axes, as shown.

Let $G(\bar{x}, \bar{y})$ be the centre of mass of the system.

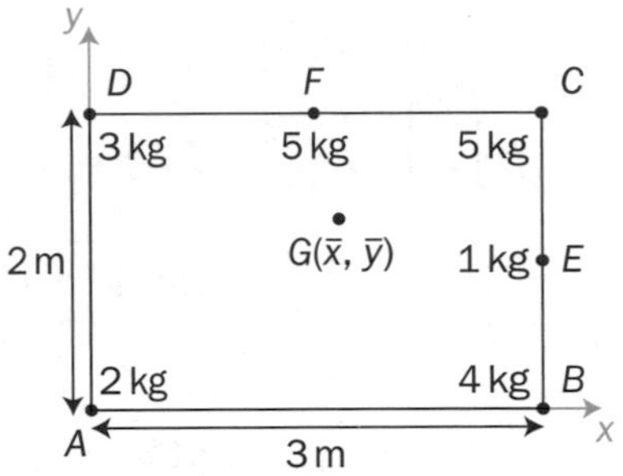

Take moments about AD:

$$\bar{x} = \frac{2\times0+4\times3+1\times3+5\times3+5\times1.5+3\times0}{2+4+1+5+5+3} = 1.875$$

Take moments about AB:

$$\bar{y} = \frac{2\times0+4\times0+1\times1+5\times2+5\times2+3\times2}{2+4+1+5+5+3} = 1.35$$

So the centre of mass is $G(1.875, 1.35)$.

Alternatively, you could work in vector form:

$$\begin{pmatrix}\bar{x}\\ \bar{y}\end{pmatrix} = \frac{2\begin{pmatrix}0\\0\end{pmatrix}+4\begin{pmatrix}3\\0\end{pmatrix}+1\begin{pmatrix}3\\1\end{pmatrix}+5\begin{pmatrix}3\\2\end{pmatrix}+5\begin{pmatrix}1.5\\2\end{pmatrix}+3\begin{pmatrix}0\\2\end{pmatrix}}{2+4+1+5+5+3} = \begin{pmatrix}1.875\\1.35\end{pmatrix}$$

For a framework of uniform heavy rods you treat each rod as a particle placed at its midpoint.

EXAMPLE 5

Three uniform rods, all of the same density, form a triangle ABC where $AB = 4$ m, $AC = 3$ m and $BC = 5$ m. Find the centre of mass of the framework.

Let the mass of 1 metre of rod be m kg.
ABC is a 3-4-5 triangle, so angle $B\hat{A}C = 90°$.
The rods are treated as particles placed at their midpoints, as shown.

Take AB and AC as the x- and y-axes:

$$\begin{pmatrix}\bar{x}\\ \bar{y}\end{pmatrix} = \frac{4m\begin{pmatrix}2\\0\end{pmatrix}+3m\begin{pmatrix}0\\1.5\end{pmatrix}+5m\begin{pmatrix}2\\1.5\end{pmatrix}}{4m+3m+5m} = \begin{pmatrix}1.5\\1\end{pmatrix}$$

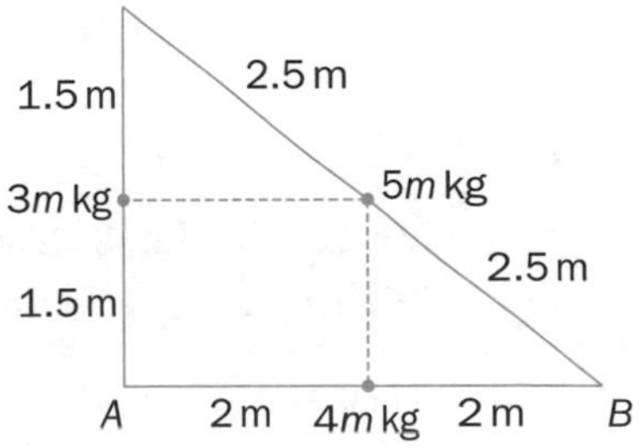

So the centre of mass is at the point $(1.5, 1)$.

Some questions require you to find an unknown mass or position needed to achieve a stated centre of mass.

M2

EXAMPLE 6

Masses of 1 kg, 3 kg and 2 kg are placed at points (2,2), (2,4) and (5,4). At what point should a mass of 4 kg be placed if the centre of mass of the system is to be at $G(3,3)$?

Let the 4 kg mass be placed at (x, y).
Using vector notation, you have

$$\begin{pmatrix}\bar{x}\\ \bar{y}\end{pmatrix} = \frac{1\begin{pmatrix}2\\2\end{pmatrix}+3\begin{pmatrix}2\\4\end{pmatrix}+2\begin{pmatrix}5\\4\end{pmatrix}+4\begin{pmatrix}x\\y\end{pmatrix}}{1+3+2+4} = \begin{pmatrix}3\\3\end{pmatrix}$$

This gives $\begin{pmatrix}18+4x\\22+4y\end{pmatrix} = \begin{pmatrix}30\\30\end{pmatrix}$ so $x = 3,\ y = 2$

So the mass should be placed at the point (3,2).

Exercise 2.1

1 Particles A, B, C and D, of mass 3 kg, 2 kg, 5 kg and 6 kg respectively, lie on a straight line with $AB = BC = CD = 1$ m. Find the distance of the centre of mass of the system from A.

2 Particles of mass 1 kg, 2 kg, 3 kg and 4 kg are attached in that order to a rod AB of length 1.5 m at distances of 0.3 m, 0.6 m, 0.9 m and 1.2 m from A.

a Assuming that the rod is of negligible mass, find the distance from A of the centre of mass of the system.

b In fact the rod is uniform and of mass m kg. The centre of mass of the system is 0.85 m from A. Find the value of m.

3 Find the coordinates of the centre of mass of the following systems of particles placed respectively at the given points.

a 3 kg, 5 kg and 7 kg at $A(2,5)$, $B(3,1)$ and $C(4,9)$

b 9 kg, 4 kg, 2 kg and 5 kg at $A(4,8)$, $B(-2,6)$, $C(4,-4)$ and $D(-2,-5)$

c 6 kg, 12 kg and 7 kg at $A(0,-8)$, $B(6,-3)$ and $C(-4,-9)$

4 Find the position vector of the centre of mass of each of the following systems.

a Masses of 3 kg, 8 kg and 5 kg at points with position vectors $3\mathbf{i} + 6\mathbf{j}$, $4\mathbf{i} - 2\mathbf{j}$ and $6\mathbf{i} - 8\mathbf{j}$ respectively.

b Masses of 3 kg, 3 kg, 4 kg and 5 kg at points with position vectors $2\mathbf{i} - \mathbf{j}$, $3\mathbf{i} + 4\mathbf{j}$, $-\mathbf{i} - 2\mathbf{j}$ and $\mathbf{i} - 3\mathbf{j}$ respectively.

5 Masses of 2 kg, 8 kg, 6 kg and 4 kg are placed respectively at the vertices A, B, C and D of a light rectangular framework $ABCD$, where $AB = 5$ cm and $BC = 3$ m. Find the distance of the centre of mass of the system from A.

6 ABC is a triangle formed of three uniform rods. AB has length 4 m and mass 4 kg. AC has length 3 m and mass 2 kg. BC has length 5 m and mass 4 kg. Find the distance of the centre of mass of the triangle from

a AB **b** AC

7 A wire, of uniform density and length 3 m, is bent to form a triangle ABC where $AB = 1.2$ m and $AC = 0.5$ m. Find the distance of the centre of mass of the triangle from

a AB **b** AC

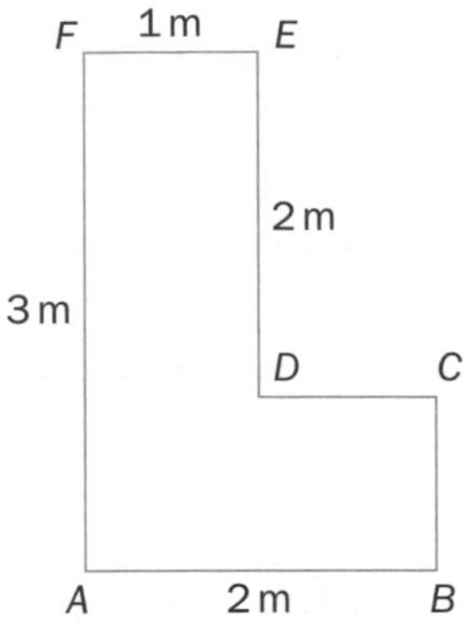

8 The diagram shows an L-shaped framework formed from a wire of uniform density. Taking AB and AF as the x- and y-axes, find the position of its centre of mass.

9 Masses of 4 kg, 9 kg and 6 kg are placed at $A(5, 3)$, $B(6, -2)$ and $C(-1, 4)$ respectively. Where should you place a particle of mass 5 kg so that the centre of mass of the whole system is $G(0, -1)$?

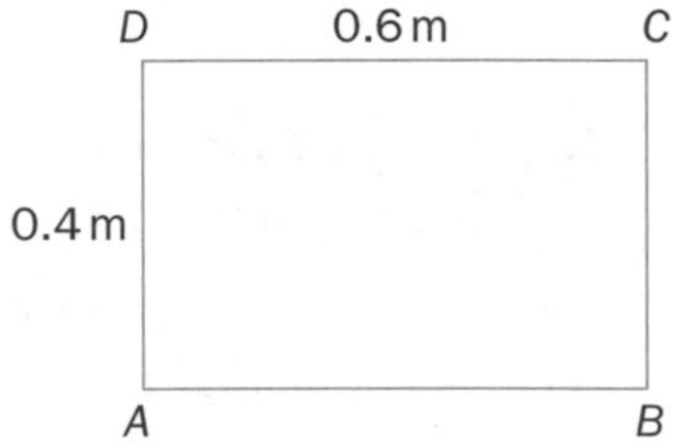

10 The diagram shows a rectangular framework $ABCD$, formed from two uniform rods each of length 1 m. The first rod is used to form ABC. The second rod has density three times greater than the first and is used to form ADC.
Taking AB and AD to be the x- and y-axes, find the position of the centre of mass of the framework.

11 Masses of 5 kg, 4 kg, 2 kg and 3 kg are placed at A, B, C and D respectively on a light rectangular framework $ABCD$, where $AB = 4$ m and $BC = 3$ m. A mass m kg is placed at a point E on CD so that the centre of mass of the system is the centre of the rectangle. Find m and the position of E.

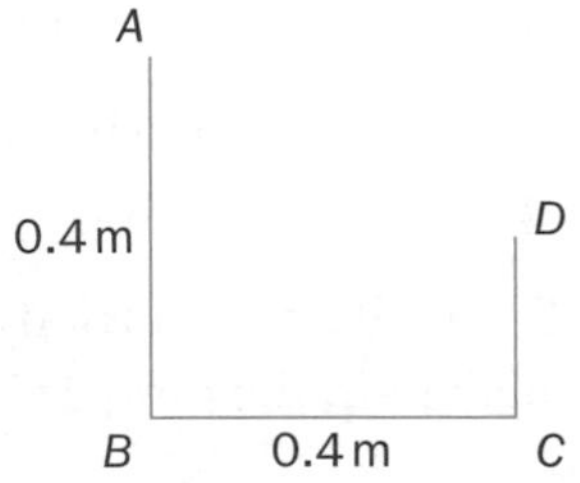

12 A uniform rod of length 1 m and mass 1 kg is bent to form the framework shown, where $\hat{B} = \hat{C} = 90°$.

a Taking BC and BA as the x- and y-axes, find the centre of mass of the framework.

b A particle of mass m kg is attached to the framework at D. The centre of mass then lies on the line AC. Find the value of m.

13 Particles of mass 3 kg, 5 kg, m_1 kg and m_2 kg are placed respectively at points (1, 4), (4, 1), (2, 1) and (4, 2). The centre of mass of the system is (3, 2).

a Show that $m_1 - m_2 = -1$, and find a second equation connecting m_1 and m_2.

b Hence find the values of m_1 and m_2.

14 Particles of mass 1 kg, 2 kg and m kg are placed respectively at points (0, 1), (1, 4) and (4, 3). The centre of mass of the system is a distance $\sqrt{13}$ from the origin. Find the value of m and the position of the centre of mass.

2.2 Centre of mass of a lamina

You can find the centre of mass of some common shapes by using their symmetry, provided that the bodies are uniformly dense.

Objects with variable density are not covered in the M2 unit.

A rectangular sheet whose thickness is negligible compared with its other dimensions is modelled as a **rectangular lamina**.

A lamina (plural *laminae*, sometimes *laminas*) is a two-dimensional figure.

By symmetry, you have:

> The centre of mass, G, of a uniform rectangular lamina $ABCD$ is at the intersection of its diagonals.

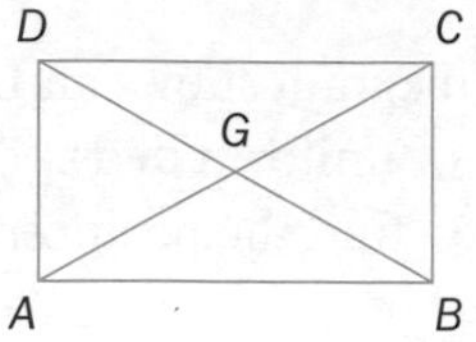

Similarly you have the centre of mass of a **circular lamina**.

> The centre of mass, G, of a uniform circular lamina is at the centre of the circle.

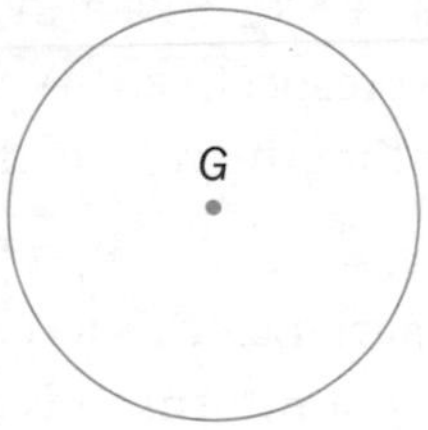

In general a **triangular lamina** has no symmetry, but there is a simple general result.

Uniform triangular lamina

You can think of a triangle as a large number of strips of negligible thickness (that is, uniform rods) parallel to one of its sides, as shown. The centres of mass (G_1, G_2, G_3, etc.) of these strips lie at their midpoints.

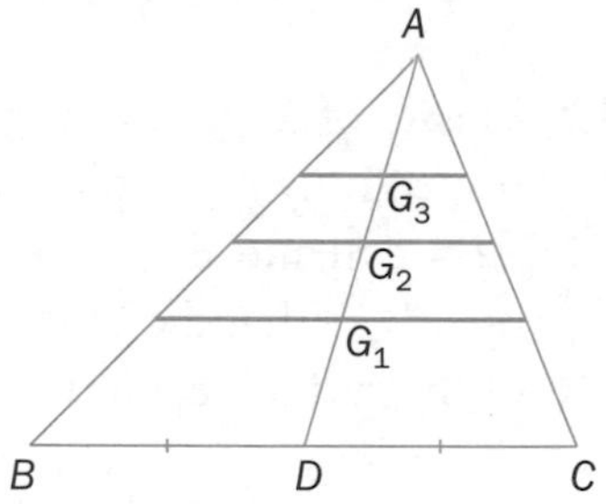

The centre of mass, G, of the triangle must lie on the line formed by G_1, G_2, G_3, etc. This is the line AD joining A to the midpoint D of BC. This line is called a **median** of the triangle.

Dividing the triangle into strips parallel to AC shows that G lies on the median BE. Similarly, G also lies on the median CF.

The medians of a triangle meet at the point which divides each median in the ratio 2 : 1. So, in this diagram, you have

$$AG : GD = BG : GE = CG : GF = 2 : 1$$

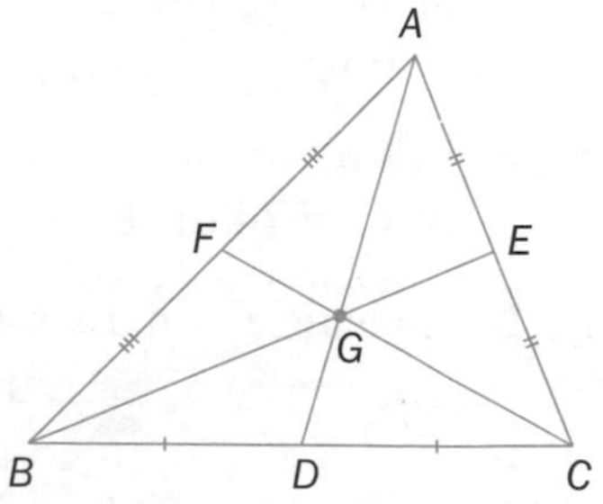

> The centre of mass, G, of a triangular lamina is at the point of intersection of its medians. G is $\frac{2}{3}$ of the way along the median from the vertex.

This formula is provided in the formulae booklet for the examination.

EXAMPLE 1

ABC is a triangle, right-angled at A and with $AB = 4.8$ m and $AC = 3.6$ m. Find the position of its centre of mass.

G is on the median BE and $EG = \frac{1}{3}EB$.

EGI and EBA are similar triangles, so $IG = \frac{1}{3}AB = 1.6$m

In the same way, G is on the median CF and $FG = \frac{1}{3}FC$.

FGH and FCA are similar triangles, so $HG = \frac{1}{3}AC = 1.2$ m

So the centre of mass is 1.2 m from AB and 1.6 m from AC.

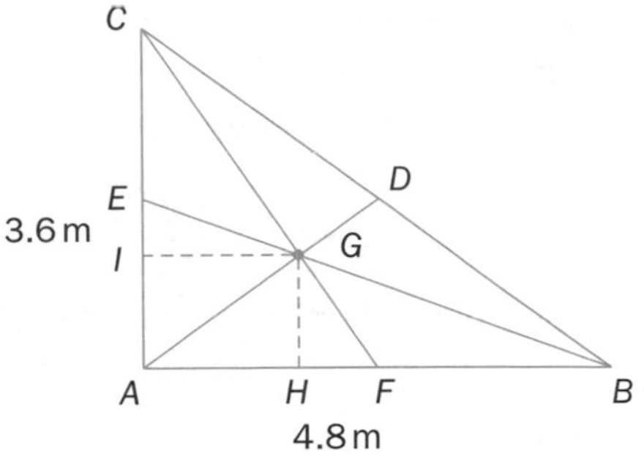

The distance of the centre of mass from a given side of a triangular lamina is $\frac{1}{3}$ the height of the triangle when that side is taken as the base.

This result was shown for a right-angled triangle in Example 1.

The other lamina you will meet requires a formula.

The diagram shows a sector of a circle of radius r.
The angle of the sector is 2α radians.

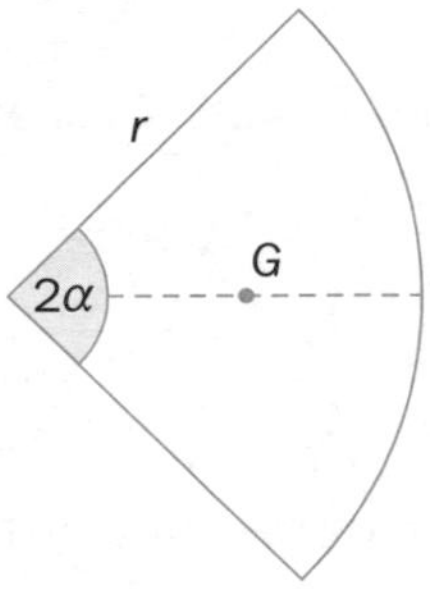

The centre of mass, G, of a sector of a circle of radius r and angle at the centre 2α radians is on the axis of symmetry at a distance $\frac{2r\sin\alpha}{3\alpha}$ from the centre.

In addition to laminae, you will meet a uniform rod bent to form a circular arc.

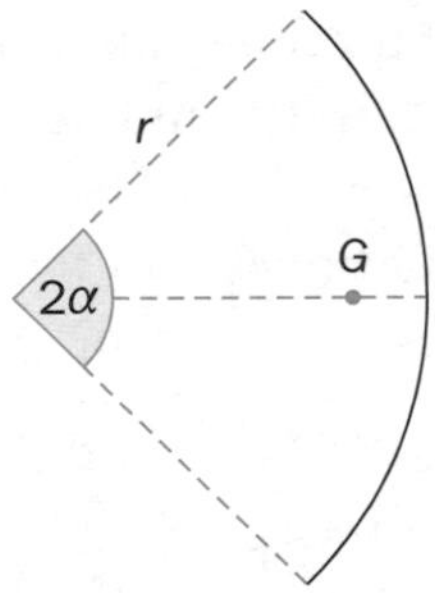

The centre of mass, G, of a uniform circular arc of radius r and angle at the centre 2α radians is on the axis of symmetry at a distance $\frac{r\sin\alpha}{\alpha}$ from the centre.

These formulae are provided in the formula booklet. Their derivations are not in the M2 syllabus. Remember to use **radians**.

M2

EXAMPLE 2

Find the centre of mass of a semicircular lamina of radius 2 m.

A semicircular lamina is a sector with an angle at the centre of π radians, so in the formula you have $\alpha = \frac{\pi}{2}$.

The centre of mass, G, is on the axis of symmetry, as shown,

and $OG = \dfrac{2 \times 2\sin\frac{\pi}{2}}{\frac{3\pi}{2}}\,\text{m} = \dfrac{8}{3\pi}\,\text{m}$

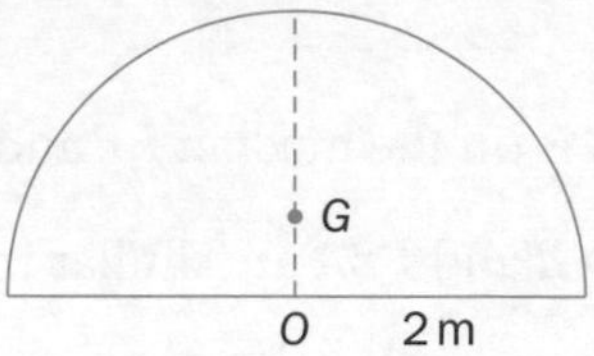

EXAMPLE 3

A uniform wire of length 2.4 m is bent to form a circular arc of radius 1.2 m. Find the position of its centre of mass.

The angle at the centre is $\frac{2.4}{1.2} = 2$ radians, so in the formula you have $\alpha = 1$ radian.

The centre of mass, G, is on the axis of symmetry, as shown,

and $OG = \dfrac{r\sin\alpha}{\alpha}\,\text{m} = \dfrac{1.2\sin 1}{1}\,\text{m} = 1.01\,\text{m}$

Use the relationship $s = r\theta$ between arc length, radius and angle (in radians).

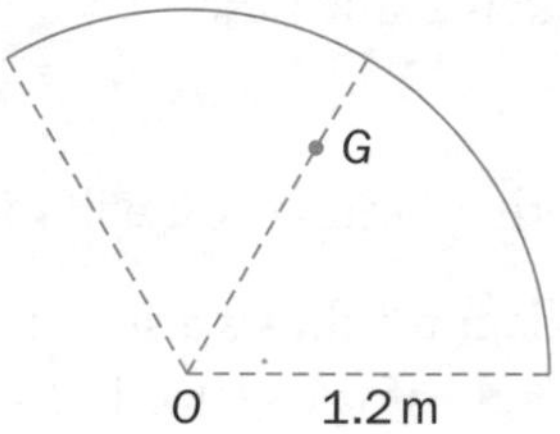

M2

You can deal with a more complex object provided it is a combination of the standard shapes. You can regard each of these components as a particle placed at its centre of mass, and find the centre of mass of the whole object from this system of particles.

EXAMPLE 4

The diagram shows an L-shaped lamina $ABCDEF$ with uniform density. Find its centre of mass.

The lamina consists of two rectangles, $AHEF$ and $HBCD$, as shown.

If the mass of 1 m^2 is m kg, then $AHEF$ is $28m$ kg and $HBCD$ is $24m$ kg.

Take AB and AF as the x- and y-axes.
The whole body behaves as particles of $28m$ kg at $G_1(2, 3.5)$ and $24m$ kg at $G_2(7, 2)$.

You have $$\begin{pmatrix}\bar{x}\\ \bar{y}\end{pmatrix} = \frac{28m\begin{pmatrix}2\\3.5\end{pmatrix} + 24m\begin{pmatrix}7\\2\end{pmatrix}}{28m + 24m} = \begin{pmatrix}4.31\\2.81\end{pmatrix}$$

So the centre of mass is $G(4.31, 2.81)$ m.

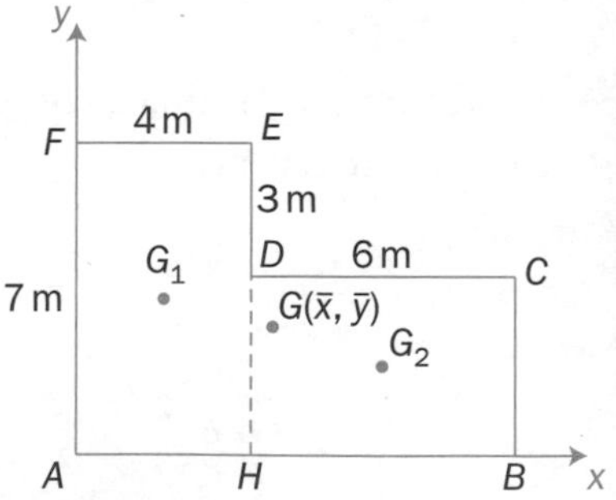

The density m cancels out. This is always true for a uniform lamina, and so you may as well assume that the density is 1. This is equivalent to using the area as the mass. Examples 5 and 6 will use this approach.

EXAMPLE 5

A uniform lamina consists of a rectangle *ABCD*, with $AB = 0.6$ m and $BC = 0.3$ m, a semicircle with *CD* as diameter and a right-angled triangle *BEC*, where $BE = 0.3$ m, as shown. Taking *AB* and *AD* as the *x*- and *y*-axes, find the centre of mass of the lamina.

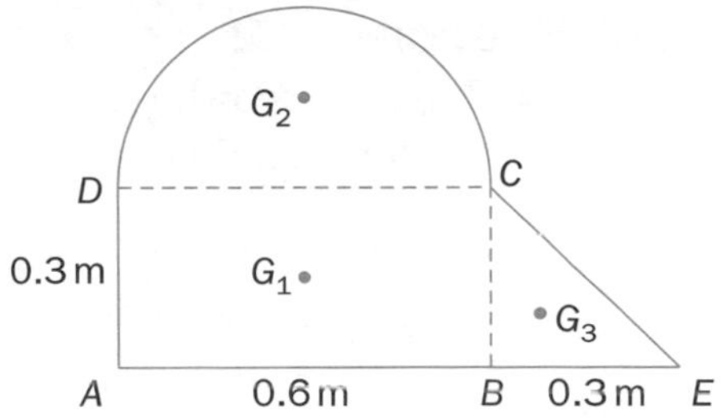

For the rectangle: area $= 0.3 \times 0.6\,\text{m}^2 = 0.18\,\text{m}^2$

centre of mass $= G_1(0.3, 0.15)$

For the semicircle: area $= \frac{1}{2}\pi \times 0.3^2\,\text{m}^2 = 0.141\,\text{m}^2$

centre of mass $= G_2\left(0.3, 0.3 + \frac{4 \times 0.3}{3\pi}\right) = G_2(0.3, 0.427)$

Centre of mass of semicircle using the formula $\frac{2r\sin\alpha}{3\alpha}$

For the triangle: area $= \frac{1}{2} \times 0.3 \times 0.3\,\text{m}^2 = 0.045\,\text{m}^2$

centre of mass $= G_3(0.7, 0.1)$

Centre of mass of triangle is $\frac{1}{3}$ of *BE* from *CB* and $\frac{1}{3}$ of *CB* from *BE*.

The whole lamina has centre of mass $G(\bar{x}, \bar{y})$ where

$$\begin{pmatrix}\bar{x}\\ \bar{y}\end{pmatrix} = \frac{0.18\begin{pmatrix}0.3\\ 0.15\end{pmatrix} + 0.141\begin{pmatrix}0.3\\ 0.427\end{pmatrix} + 0.045\begin{pmatrix}0.7\\ 0.1\end{pmatrix}}{0.18 + 0.141 + 0.045} = \begin{pmatrix}0.349\\ 0.251\end{pmatrix}$$

The density has been taken as 1, so the area represents the mass.

So the centre of mass is $G(0.349, 0.251)$ m.

In some cases the figure you are working with can best be regarded as a standard shape with one or more pieces removed.

EXAMPLE 6

The diagram shows a uniform rectangular piece of card *ABCD* from which a rectangle *PQRS* has been cut. Taking *AB* and *AD* as the *x*- and *y*-axes, find the centre of mass of the remaining lamina.

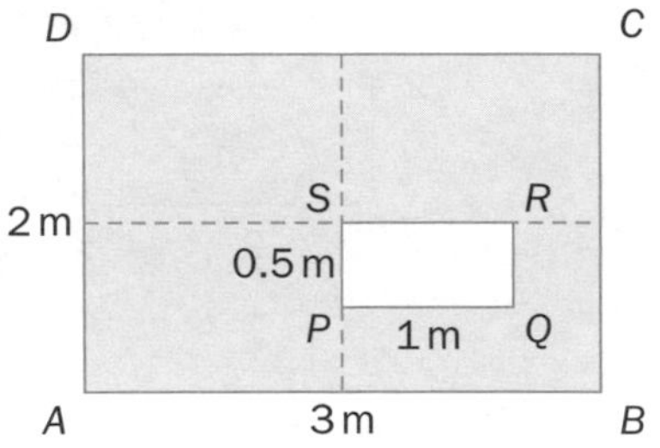

For the original rectangle: area $= 6\,\text{m}^2$, centre of mass is $(1.5, 1)$

For the piece removed: area $= 0.5\,\text{m}^2$, centre of mass is $(2, 0.75)$

For the remaining lamina: area $= 5.5\,\text{m}^2$, centre of mass is $(\bar{x}, \bar{y})$

Take moments about the axes:

Moment of *ABCD* = moment of shaded lamina + moment of *PQRS*

$$6\begin{pmatrix}1.5\\ 1\end{pmatrix} = 5.5\begin{pmatrix}\bar{x}\\ \bar{y}\end{pmatrix} + 0.5\begin{pmatrix}2\\ 0.75\end{pmatrix}$$

and so $\quad 5.5\begin{pmatrix}\bar{x}\\ \bar{y}\end{pmatrix} = 6\begin{pmatrix}1.5\\ 1\end{pmatrix} - 0.5\begin{pmatrix}2\\ 0.75\end{pmatrix} = \begin{pmatrix}8\\ 5.625\end{pmatrix}$

which gives $\quad \begin{pmatrix}\bar{x}\\ \bar{y}\end{pmatrix} = \begin{pmatrix}1.45\\ 1.02\end{pmatrix}$

So the centre of mass of the shaded lamina is $(1.45, 1.02)$.

M2

Exercise 2.2

1 Find the coordinates of the centre of mass of each of the following uniform laminae.

a

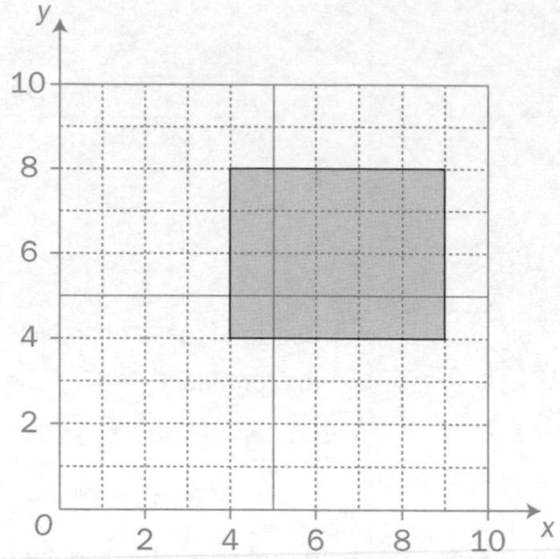

b

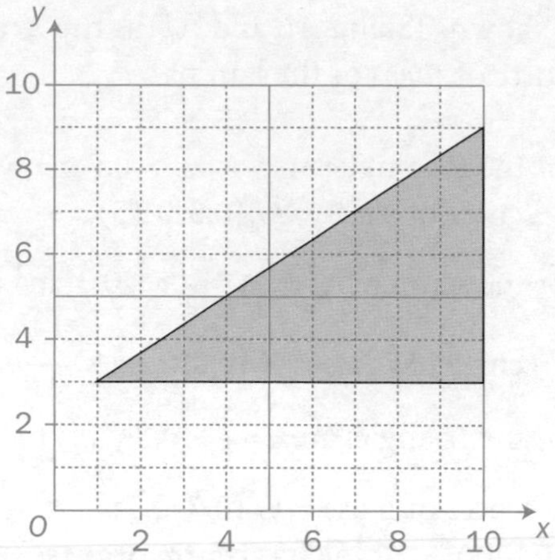

c

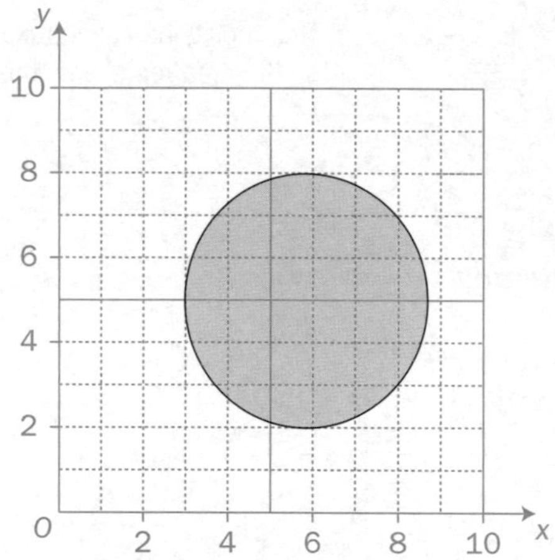

d

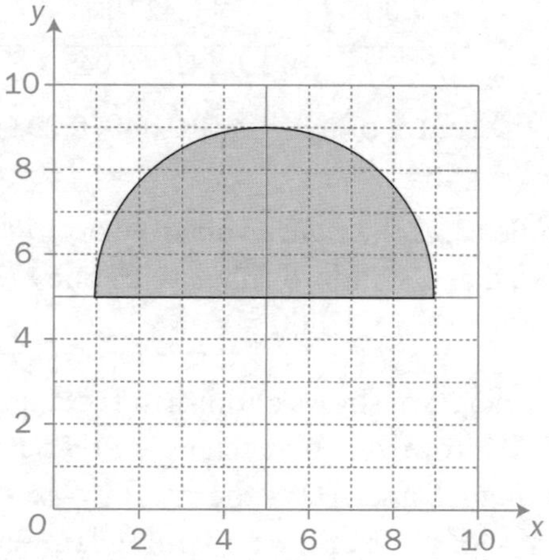

e

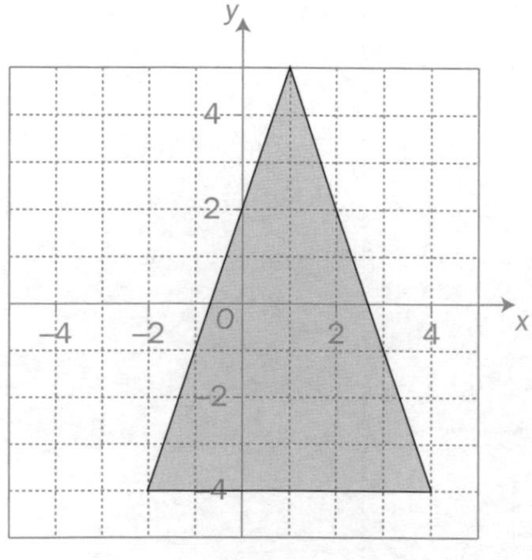

f

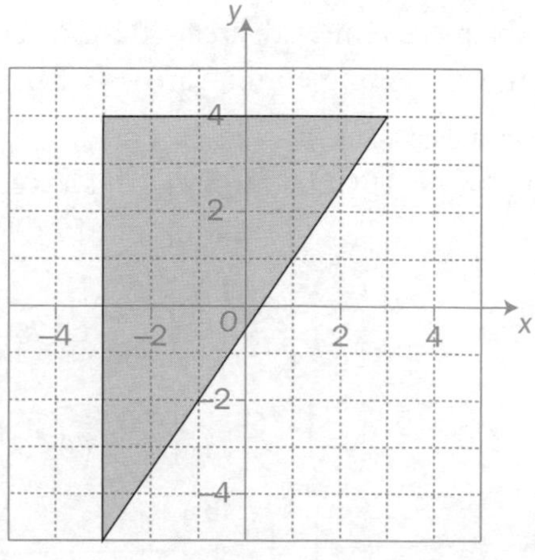

2 Find the coordinates of the centre of mass of each of the following composite uniform laminae.

a

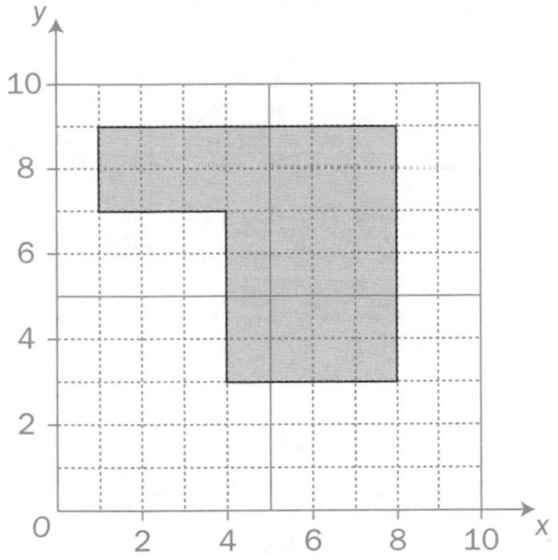

b

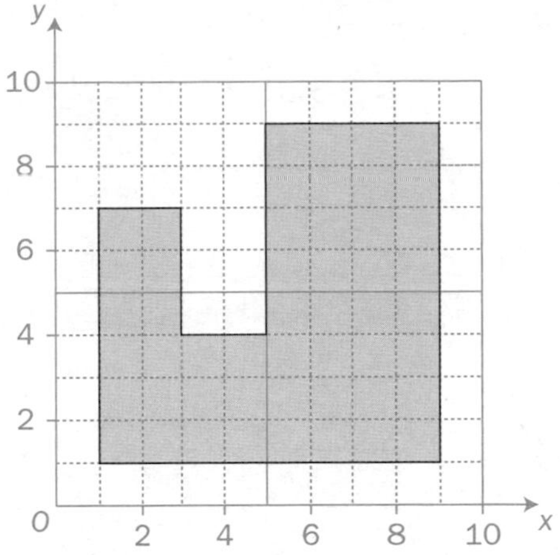

c

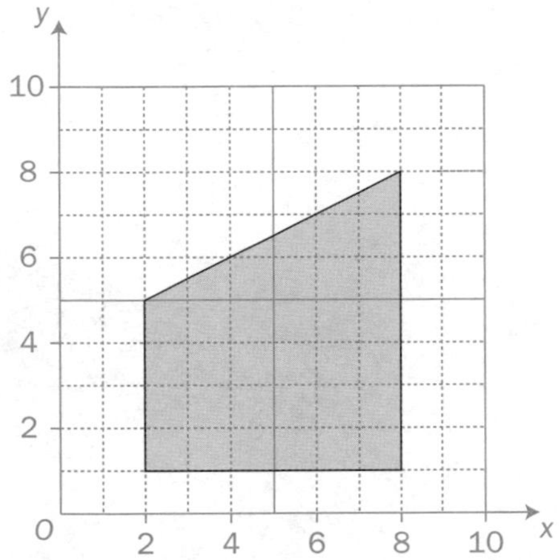

d

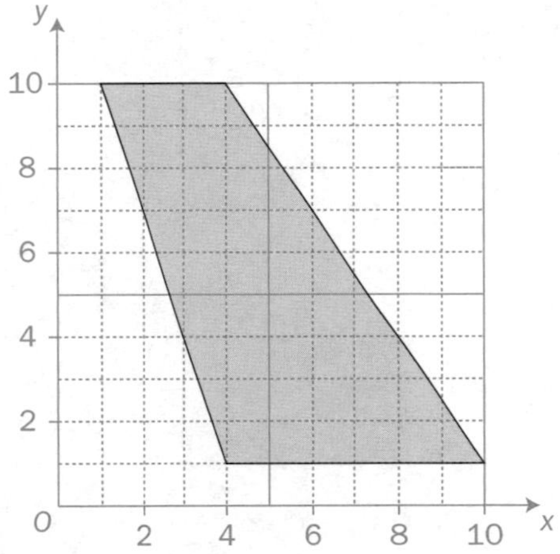

e

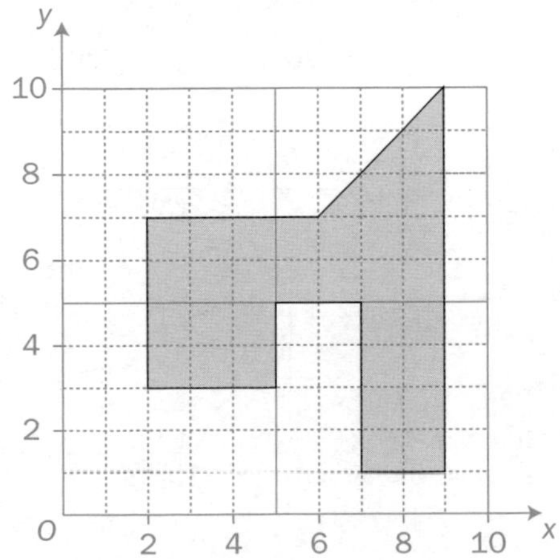

f

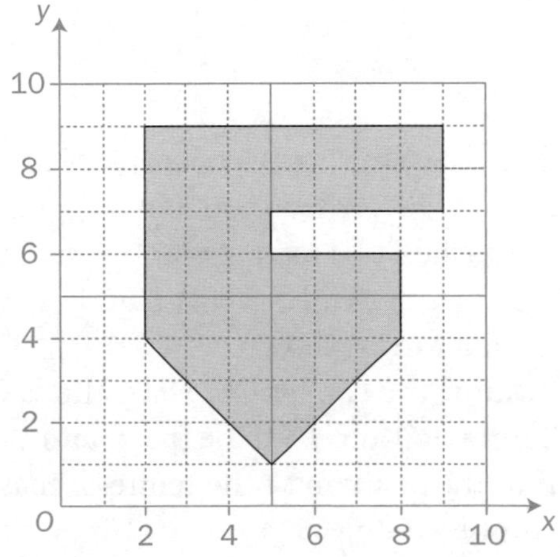

3 Find the centre of mass of each of the following uniform laminae relative to the origin O and the axes shown.

a

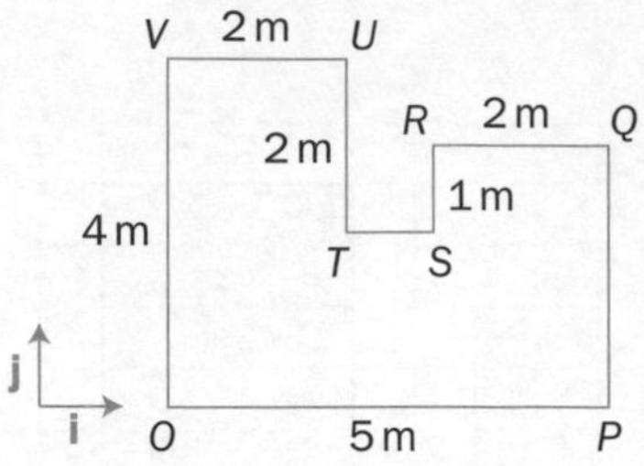

b

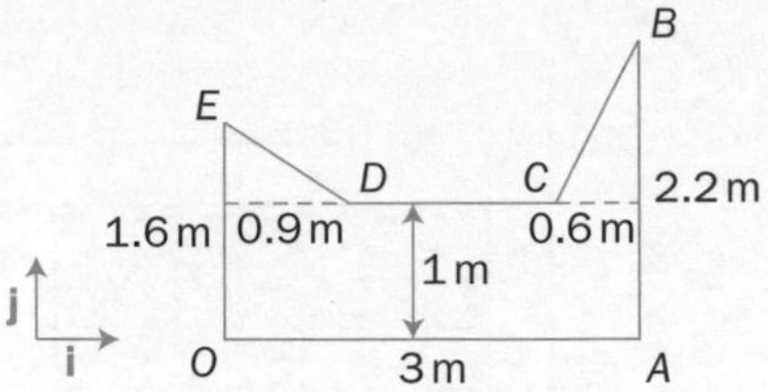

c

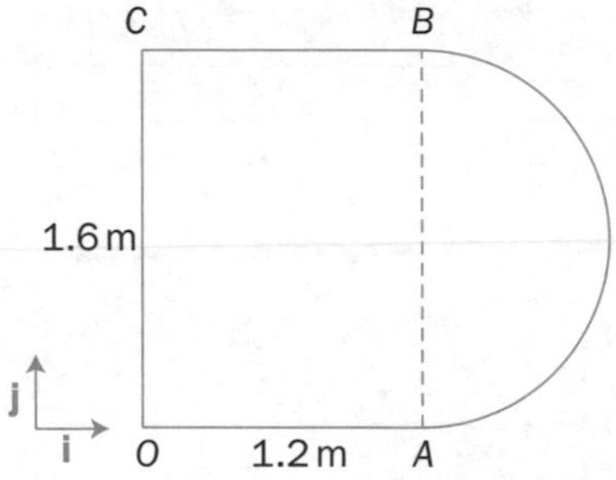

d

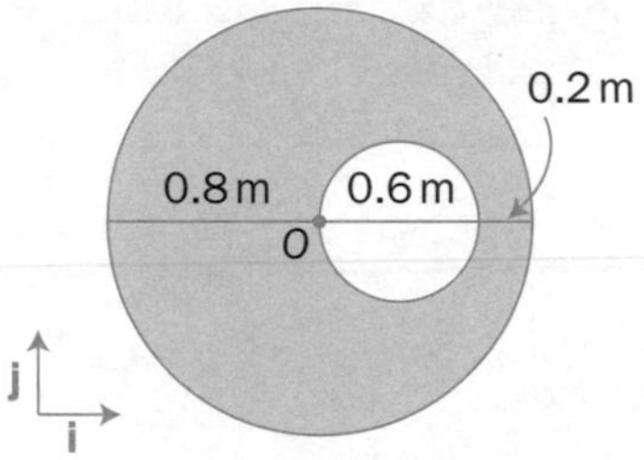

e

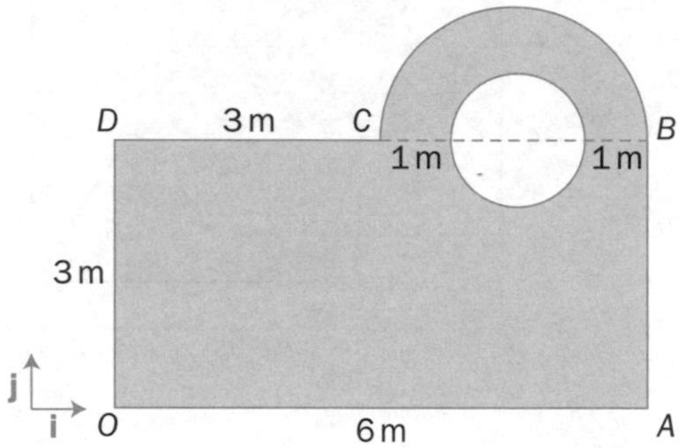

f

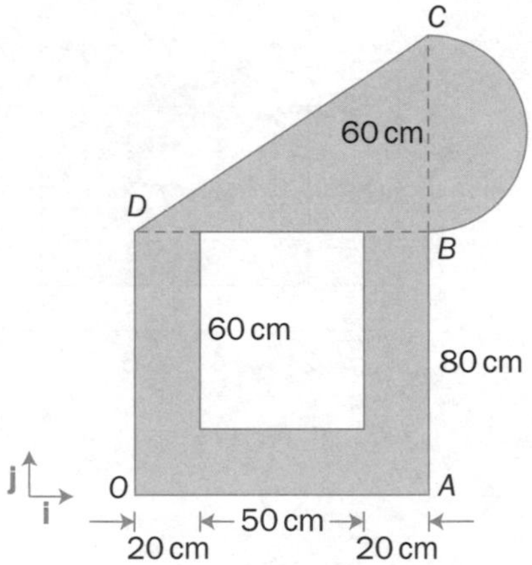

4 The diagram shows a triangular lamina ABC, with $A\hat{B}C = 90°$, $AB = 0.6$ m and $BC = 0.9$ m. The triangle is attached to a rectangular lamina $PQRS$, where $PQ = 1.2$ m and $PS = 0.8$ m. BC lies on PQ and $PB = 0.2$ m, as shown.
Assume the two laminae have the same uniform density. Taking PQ and PS to be the x- and y-axes respectively, find the position of the centre of mass of the combined object.

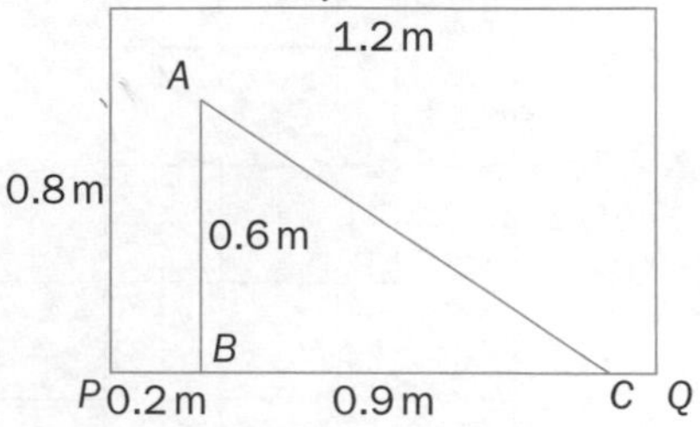

5 A uniform rectangular card $ABCD$ is folded along OF and BE as shown in the diagram.
$AB = 120$ cm, $AD = 40$ cm, $AO = 20$ cm and $CE = 40$ cm.
Taking O as the origin and axes as shown, find the centre of mass of the folded card.

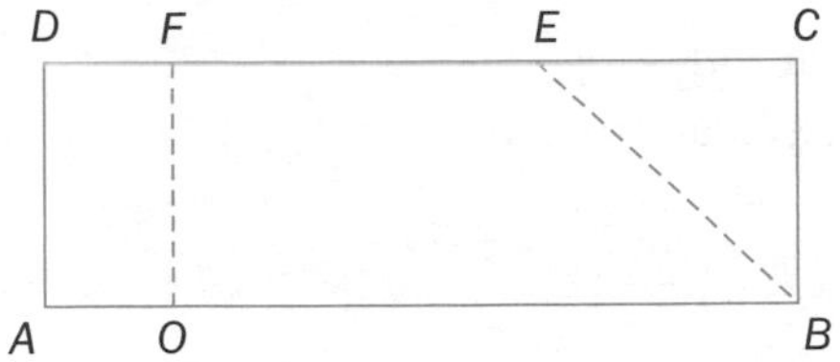

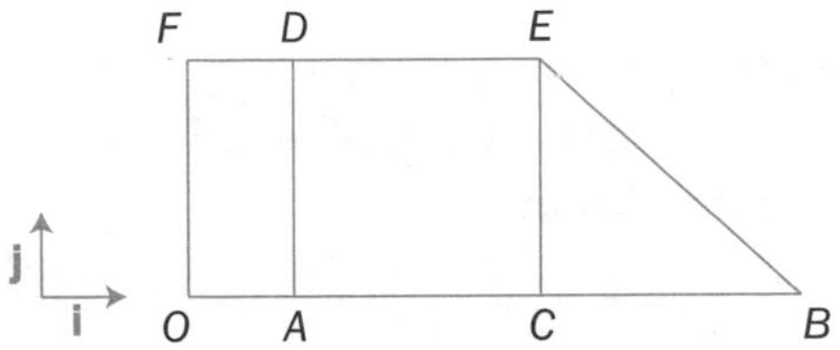

6 The diagram shows a lamina in the form of a sector of a circle centre O of radius 6 cm. The angle of the sector is 60°.
A uniform rod of negligible width is bent so that it attaches exactly to the arc of the sector, as shown.

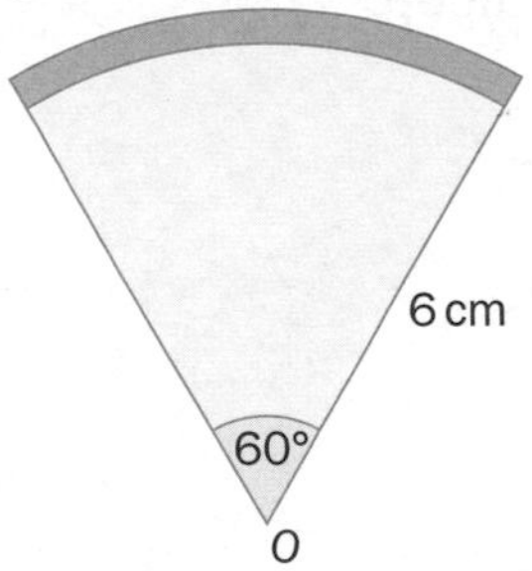

The mass of the lamina is twice the mass of the rod. Show that the centre of mass of the combined object is $\frac{14}{\pi}$ cm from O.

7 A bridge on a model railway is in the form of a rectangle with a semicircular arch cut out of it. The whole structure is 28 cm high and 36 cm wide, and the radius of the arch is 10 cm.

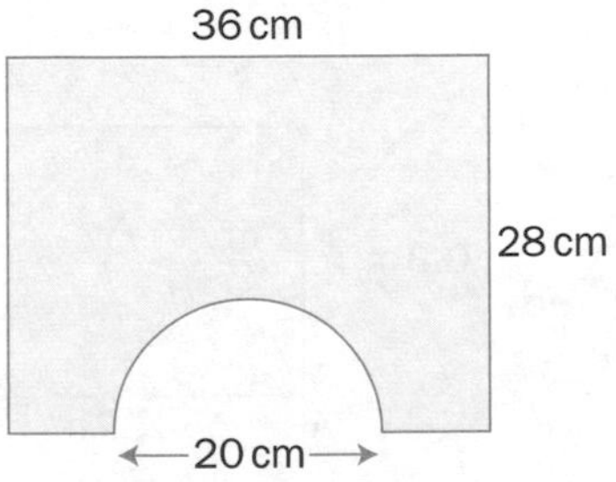

Assuming that the bridge is made of a uniformly dense material, find the height of its centre of mass above the base.

M2

8 A uniform circular disc has centre O and radius 6 cm. Two circular holes, each of radius 1 cm, are cut from the disc. The centres of the holes are at A and B, where $OA = OB = 3$ cm and angle $A\hat{O}B = 90°$. Find the position of the centre of mass of the remainder.

9 A basketball hoop is made from a single uniform metal rod. It comprises a straight section of length 20 cm and a circular hoop of diameter 30 cm.

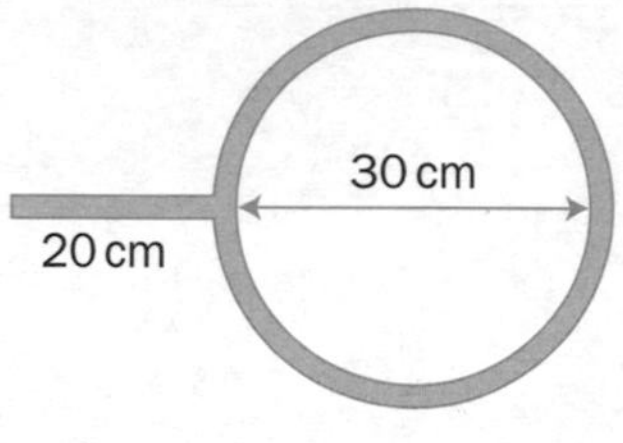

Find the distance of its centre of mass from the centre of the hoop.

10 A uniform lamina is in the shape of an equilateral triangle ABC of side length 2 m. E and F are the midpoints of BC and AC. The triangle EFC is removed. Find the exact distance from AB of the centre of mass of the remaining trapezium $ABEF$.

11 A uniform lamina is in the form of a segment of a circle of radius 60 cm cut off by a chord AB of length 60 cm. Find the distance of the centre of mass of the segment from AB.

12 A model aircraft wing comprises fabric attached to a triangular framework ABC. $AB = 80$ cm and $AC = BC = 50$ cm. Each centimetre of the framework has mass 2 g, and each square centimetre of fabric has mass 0.1 g. Find the distance of the centre of mass of the wing from AB.

13 The diagram shows a lamina comprising a rectangle $ABDE$ of length 1.2 m and width 0.8 m together with an isosceles triangle BCD of height 0.6 m. F is the midpoint of AE. H is a point on CF, with $FH = 0.4$ m. A hole, centre H and radius r cm, is cut in the lamina. The centre of mass of the object is at G.

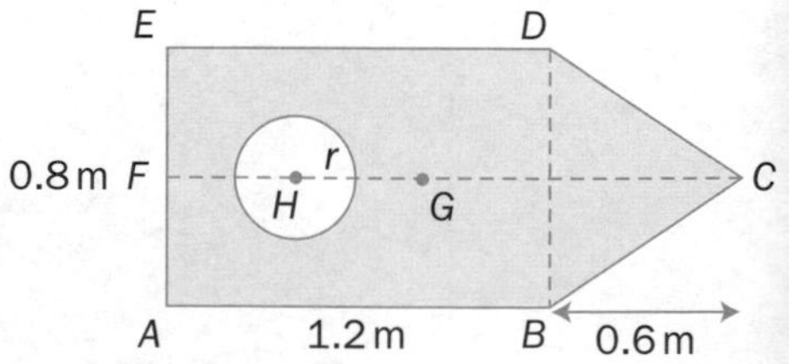

a Explain why G lies on CF.

b If $r = 0.2$, find the distance FG.

c If $FG = 1$ m, find the value of r.

d Explain why FG cannot be 1.1 m.

14 A uniform wire of length 1.4 m is bent to form the arc and two radii of a sector of a circle of radius 0.5 m.
Find the distance of the centre of mass from the centre of the circle.

15 The diagram shows a carpenter's square, with a rectangular handle *PQUV*, of length 0.2 m and width 0.1 m, and a rectangular blade *RSTU* of length 0.3 m and width 0.1 m. The blade is made from metal of density 1 kg m^{-2}, and the handle from material of density m kg m^{-2}.

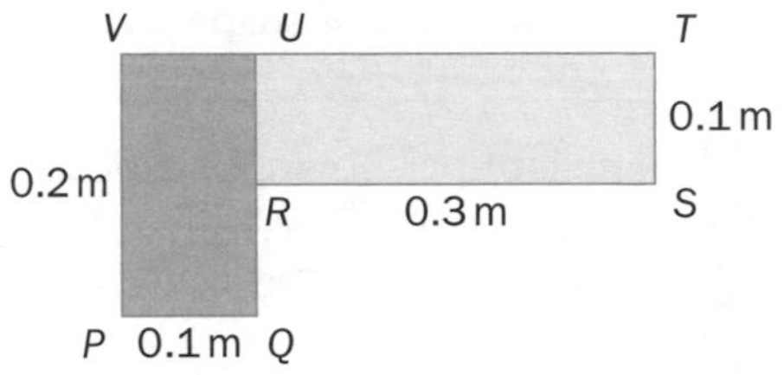

a Find the value of m if the centre of mass of the object lies on the line *QU*.

b For this value of m, find the distance of the centre of mass from *PQ*.

16 An isosceles trapezium *ABCD* has parallel sides *AB* and *CD* of lengths a and b respectively and height h.

a Prove that the centre of mass of *ABCD* is at a distance

$$\frac{h(a+2b)}{3(a+b)}$$

from *AB*.

b Was it necessary for *ABCD* to be isosceles for this result to hold? Explain your answer.

17 The diagram shows a lamina in the form of an isosceles triangle *ABC*, where $AC = BC$, of height h. *D* lies on *BC* and $BD:DC = 1:2$. *E* lies on *AC* and $AE:EC = 1:2$. The portion of the triangle above the line *DE* has twice the density of the remainder. Find the distance of the centre of mass of *ABC* from *AB*.

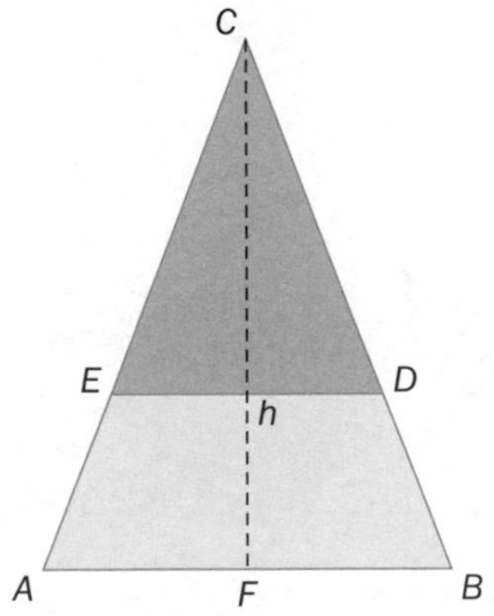

2.3 Centre of mass and equilibrium

If an object is suspended from a single point, its equilibrium position depends on its centre of mass. The centre of mass must be directly below the point of suspension.

EXAMPLE 1

The lamina shown is suspended freely from the vertex A. Find the angle between AB and the vertical when the lamina is in equilibrium.

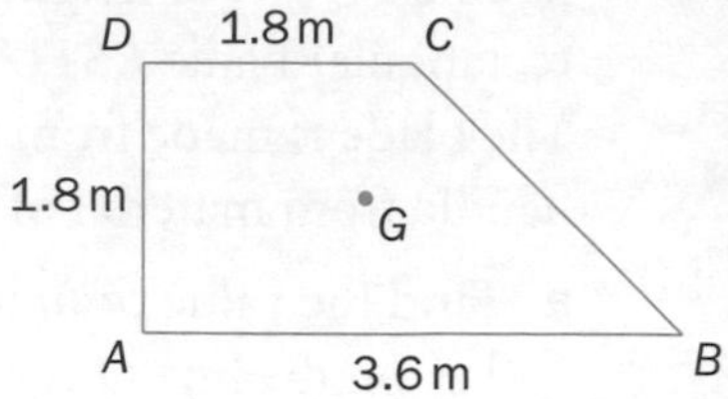

Take AB and AD as the x- and y-axes. The centre of mass of the lamina is $G(\bar{x}, \bar{y})$.

The lamina comprises:

a square of area 3.24 m^2, centre of mass (0.9, 0.9)

and a triangle of area 1.62 m^2, centre of mass (2.4, 0.6)

You have $\begin{pmatrix}\bar{x}\\ \bar{y}\end{pmatrix} = \dfrac{3.24\begin{pmatrix}0.9\\0.9\end{pmatrix}+1.62\begin{pmatrix}2.4\\0.6\end{pmatrix}}{3.24+1.62} = \begin{pmatrix}1.4\\0.8\end{pmatrix}$

So the centre of mass $= G(1.4, 0.8)$

When the lamina is suspended from A, the line AG is vertical. AB makes the angle θ with the vertical, as shown.

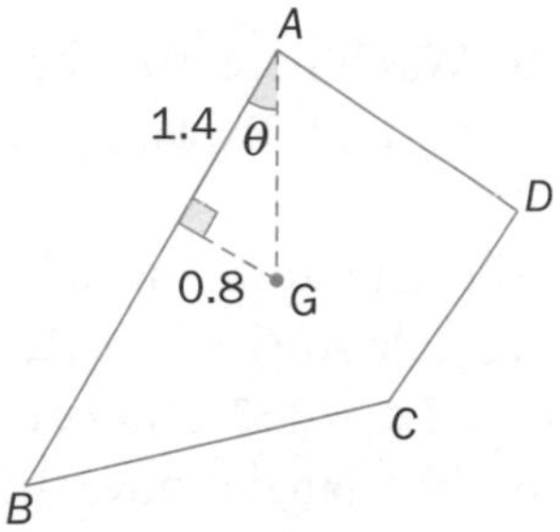

You have $\tan\theta = \dfrac{0.8}{1.4}$

and so $\theta = 29.7°$

The centre of mass also affects equilibrium when an object is placed on an inclined plane. If the angle of the slope is too great, the object can tip over.

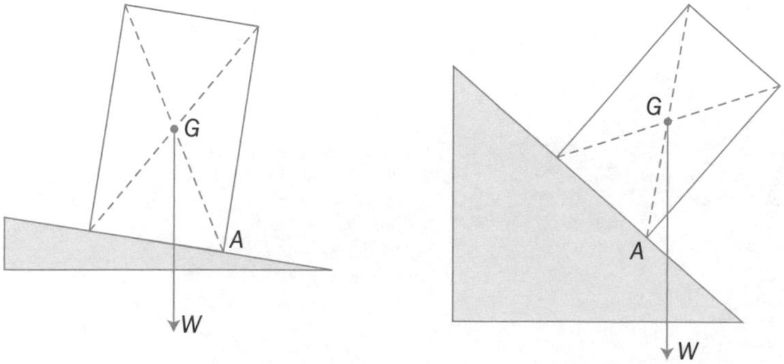

When this rectangle is placed on a slight incline, its weight has an anticlockwise moment about A, so the rectangle can rest in equilibrium. On the steeper slope the moment is clockwise, causing the rectangle to tip over.

This assumes that friction is sufficient to prevent the block from sliding before it reaches the point at which it tips over.

EXAMPLE 2

The lamina in Example 1 is placed with CD on the line of greatest slope of a plane inclined at α to the horizontal, D being higher up the slope than C. Find the greatest value of α for which the lamina will remain in equilibrium.

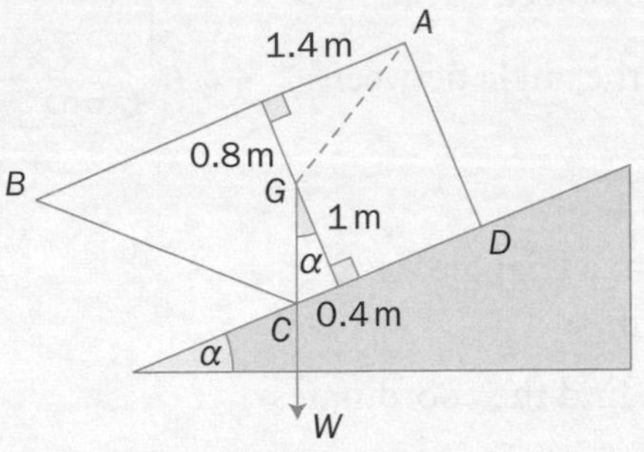

The diagram shows the lamina on the point of tipping. You have

$$\tan\alpha = \frac{0.4}{1}$$

and so $\alpha = 21.8°$

The lamina rests in equilibrium provided the slope does not exceed 21.8°.

If α were smaller, the weight would act to the right of C, maintaining equilibrium. If α were greater, the weight would act to the left of C, producing a moment which would make the lamina tip over.

Exercise 2.3

1 The diagram shows an L-shaped uniform lamina $ABCDEF$, with $AB = 4$ m, $BC = 1$ m, $EF = 2$ m and $AF = 3$ m.

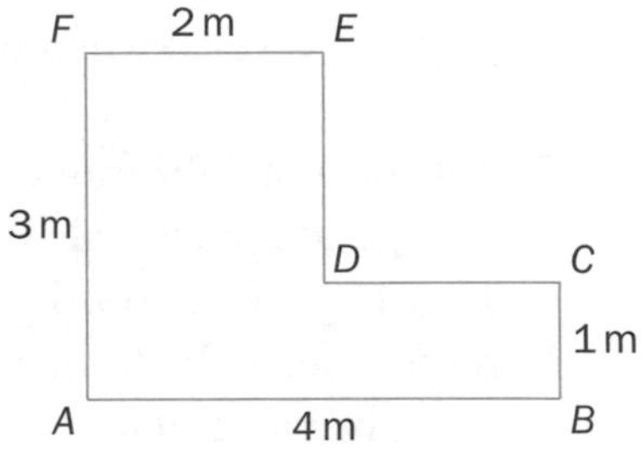

a Find the distance of the centre of mass of the lamina from

i AF

ii AB

b The lamina is attached to a hook at A and allowed to hang freely. Find the angle with the vertical made by the side AF.

c The lamina is placed on an inclined plane with EF in contact with the slope, as shown. Find the angle of the slope if the lamina is on the point of tipping over.

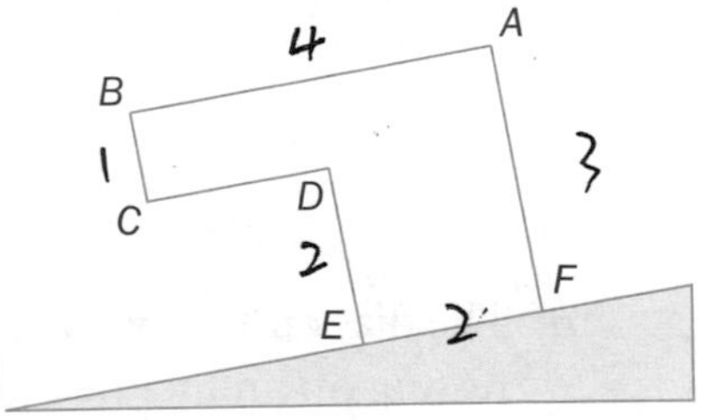

2 $ABCD$ is a uniform rectangular lamina with $AB = 60$ cm and $BC = 30$ cm. E is the midpoint of CD. The triangle BCE is removed from the lamina, and the remainder is suspended from E. Find the angle that AD makes with the vertical when the lamina hangs in equilibrium.

3 A uniform rod AB of mass 5 kg and length 2 m is attached at a point B on the rim of a uniform disc, centre C, of radius 0.6 m and mass 10 kg, so that the rod is perpendicular to the plane of the disc, as shown.

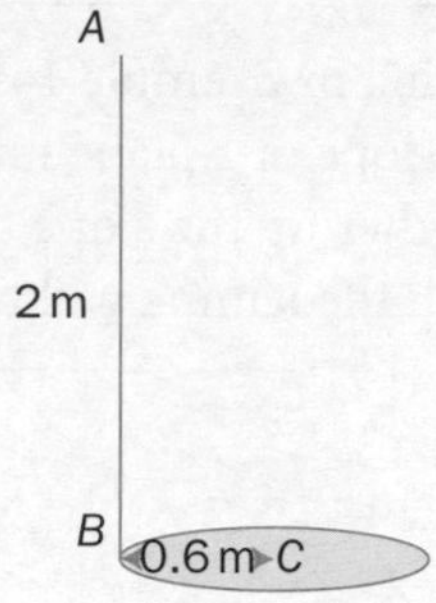

a Taking BC and BA as the x- and y-axes respectively, find the position of the centre of mass of the object.

b If the object is suspended from A, find the angle between the rod AB and the vertical.

4 A uniform rod of length 3 m is bent to form a triangle ABC, with $B\hat{A}C = 90°$, $AB = 1.2$ m and $AC = 0.5$ m.

a Taking AB and AC as the x- and y-axes, find the coordinates of the centre of mass of the triangle.

b The triangle is freely suspended from B so that it hangs in equilibrium. Find the angle between the side AB and the vertical.

5 A uniform wire AB is bent to form the arc of a semicircle. The wire is then suspended freely from A. Find the angle between the diameter AB and the vertical.

6 Masses of 2 kg, 4 kg, 6 kg and 9 kg are placed respectively at the vertices A, B, C and D of a light rectangular framework $ABCD$, where $AB = 5$ cm and $BC = 3$ m. Find the angle which AB makes with the vertical when the framework is suspended from A.

7 $ABCDE$ is a light framework consisting of a square $ABCE$ and an equilateral triangle CDE as shown. Particles of mass 2 kg, 1 kg, 4 kg, 5 kg and m kg are attached to A, B, C, D and E respectively. The framework is then suspended from A. Find the value of m if the diagonal AC makes an angle of 20° with the vertical.

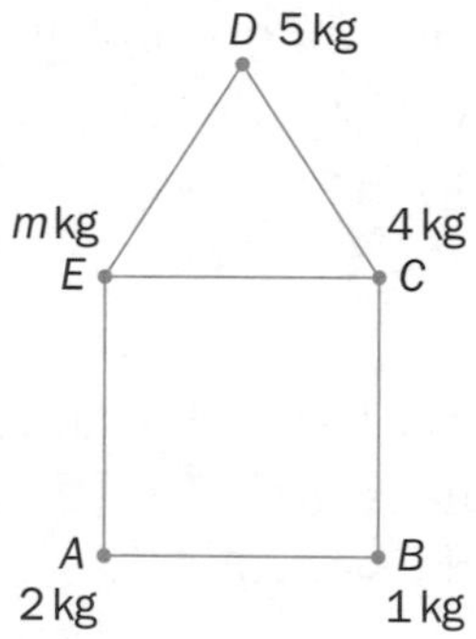

8 The diagram shows a symmetrical T-shaped uniform lamina resting on an inclined plane.
$PQ = 20$ cm, $PW = 12$ cm, $RS = 4$ cm and $ST = 20$ cm.
Find the greatest angle at which the plane can be inclined to the horizontal without the lamina tipping over.

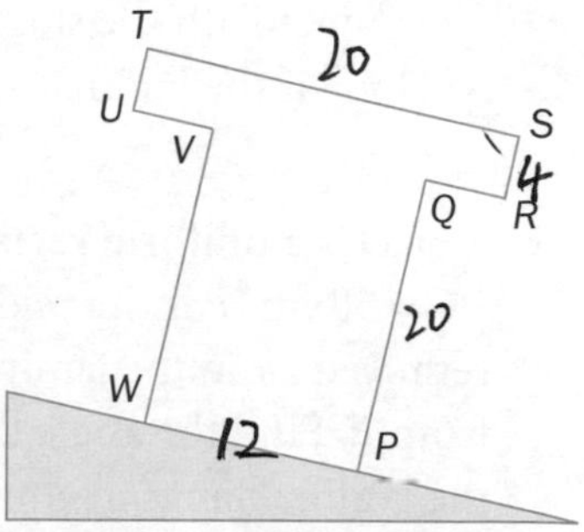

9 The diagram shows the cross-section of a prism consisting of a rectangle *ABCD*, of length 50 cm and width 30 cm, from which a triangle *BCE* has been removed so that *AB* has length x cm. Find the minimum value of x for which the prism will stand on a horizontal surface with *AE* in contact with the surface.

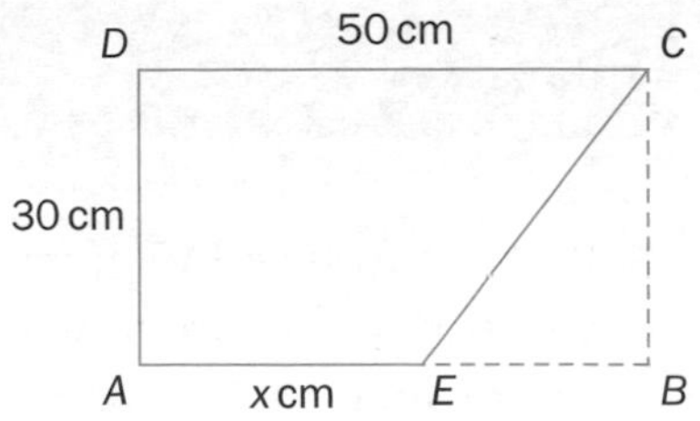

10 A uniform lamina is in the form of an isosceles trapezium of height 6 cm and with parallel sides of length 4 cm and 10 cm, as shown.

a Using the axes illustrated, find the coordinates of the centre of mass of the lamina.

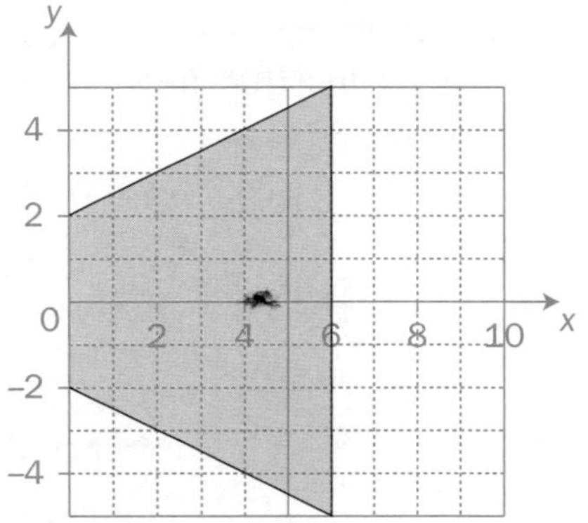

b The lamina is placed with its shortest side on an inclined plane, as shown. The lamina is on the point of toppling. Find the angle of the slope.

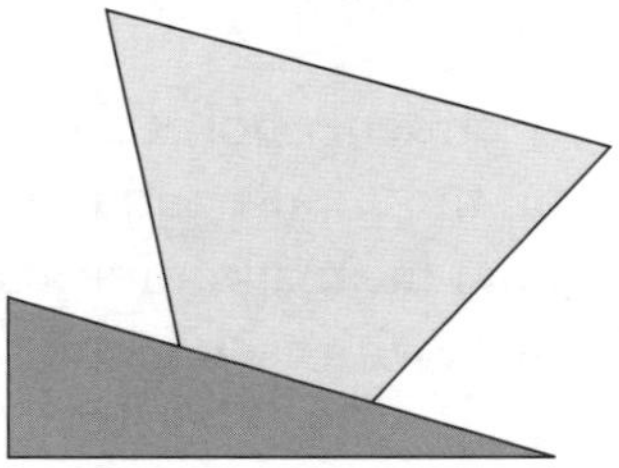

11 A uniform prism has cross-section *ABCD* comprising a square and a triangle, as shown. The prism rests on a rough plank. The coefficient of friction between the prism and the plank is 0.4. The angle, α, between the plank and the horizontal is gradually increased.

a Assuming that the prism does not topple over, what is the value of α for which the prism is on the point of sliding?

b Assuming that the prism does not slide, what is the value of α for which the prism is on the point of toppling over?

c What do you conclude about the way in which equilibrium will be broken?

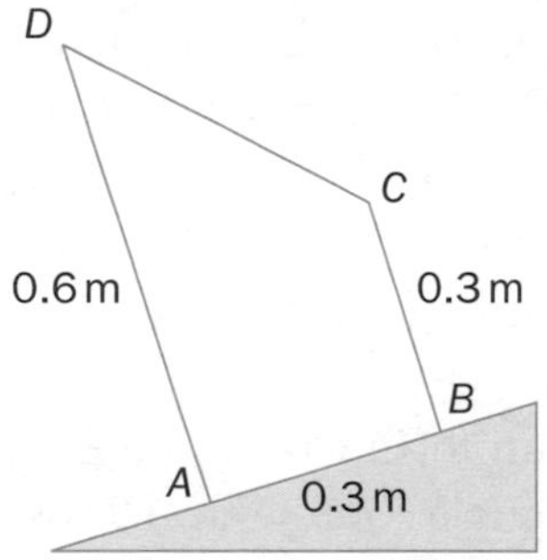

M2

1 Particles A, B and C, of mass 3 kg, 2 kg and 5 kg respectively, are attached in that order to a straight light rod with $AB = 2$ m and $BC = 3$ m. If the centre of mass of the system is G, find the length AG.

2 A lawn edging tool comprises a handle, of mass 30 g and negligible thickness, a uniform narrow shaft of length 1 m and mass 150 g, and a semicircular blade of diameter 20 cm and density 3 g cm^{-2}, as shown.

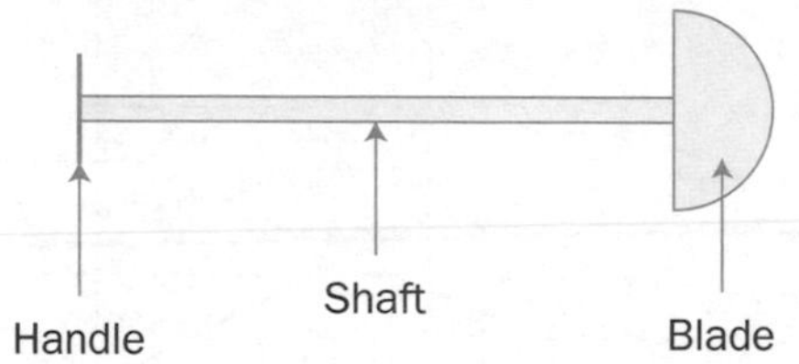

a Find the distance of the centre of mass of the tool from the handle.

b A gardener holds the tool horizontally, with one hand on the handle and the other at the midpoint of the shaft. Find the magnitude and direction of the force she must exert with each hand.

3 Particles A, B, C and D, of mass 2 kg, 3 kg, 1 kg and 4 kg respectively, are placed on a plane at points (2, 1), (7, 5), (4, −2) and (3, 8) respectively. Find the coordinates of G, the centre of mass of the system.

4 $ABCD$ is a light rectangular framework with $AB = 3$ m and $BC = 2$ m. E and F are the midpoints of BC and CD respectively. Particles of mass 3 kg, 2 kg, 1 kg, 5 kg, 6 kg and 3 kg are placed respectively at A, B, C, D, E and F. Find the distance of the centre of mass of the system from AB and AD.

5 A uniform wire of length 40 cm is bent to form a right-angled triangle ABC, with $AB = 8$ cm and $AC = 15$ cm.

a Find the distance of the centre of mass of the triangle from AB and AC.

b The triangle is suspended freely from B and hangs in equilibrium. Find the angle between the side AB and the vertical.

6 The diagram shows a uniform lamina. All the vertices are right angles.

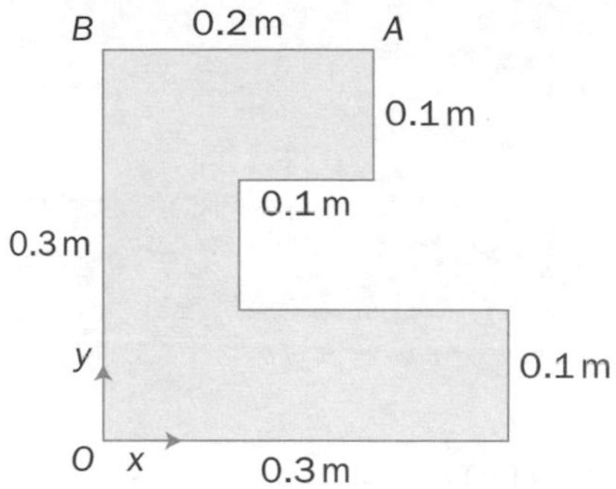

a Find the position of *G*, the centre of mass of the lamina, relative to the origin *O* and axes shown.

b The lamina is suspended freely from the vertex *A* and hangs in equilibrium. Find the angle between the side *OB* and the vertical.

7 A uniform wire *AB* of length 60 cm is bent to form an arc of a circle of radius 50 cm.

a Find the distance between the centre of mass of the arc and the chord *AB*.

A lamina is in the form of a sector of a circle with radius 50 cm and arc length 60 cm. The wire is attached along the arc of the sector.

b If the mass of the wire is 1 kg and the centre of mass of the combined object lies on the chord *AB*, find the mass of the lamina.

8 The diagram shows a lamina comprising a square of side length 0.9 m attached to a right-angled triangle. The lamina rests in equilibrium on an inclined plane, as shown. If the lamina is on the point of toppling over, find the angle of the slope.

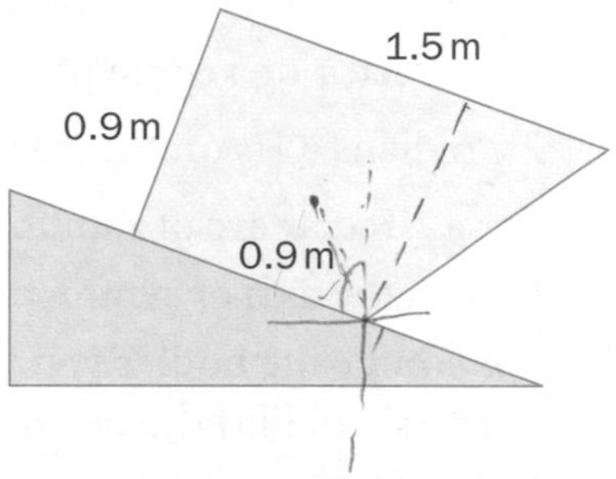

9 A uniform lamina consists of a semicircle with centre *O* and radius r cm attached to an isosceles triangle of base $2r$ cm and height h cm, as shown.

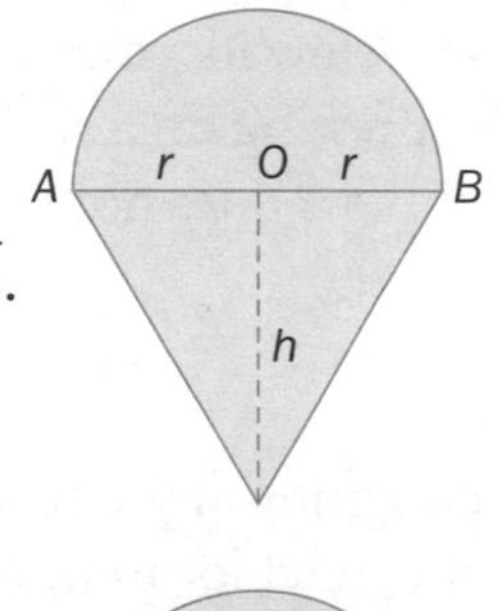

a If the centre of mass of the lamina is at *O*, show that $h = r\sqrt{2}$.

In fact $r = 15$ and $h = 30$. A hole of radius R cm is cut in the triangle, as shown, with centre C on the median of the triangle, where $OC = 10$ cm.

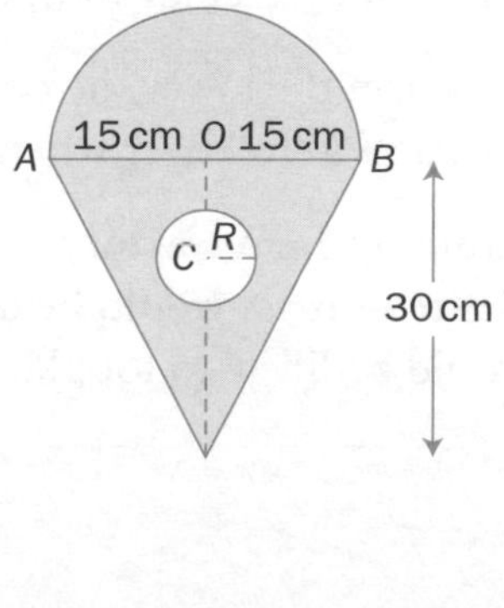

b If the centre of mass of this lamina is at *O*, find the value of *R*.

2 Exit

Summary

Refer to

- For particles of mass $m_1, m_2, \ldots, m_n$ placed at points with coordinates $(x_1, y_1), (x_1, y_1), \ldots, (x_n, y_n)$, the centre of mass is $G(\bar{x}, \bar{y})$, where

$$\bar{x} = \frac{m_1x_1 + m_2x_2 + \cdots + m_nx_n}{m_1 + m_2 + \cdots + m_n} = \frac{\sum_{i=1}^{n} m_ix_i}{\sum_{i=1}^{n} m_i}$$

$$\bar{y} = \frac{m_1y_1 + m_2y_2 + \cdots + m_ny_n}{m_1 + m_2 + \cdots + m_n} = \frac{\sum_{i=1}^{n} m_iy_i}{\sum_{i=1}^{n} m_i}$$ 2.1

- You can find the centres of mass of some common shapes by symmetry.
- For a triangular lamina, G is at the point of intersection of the medians. G is $\frac{2}{3}$ of the way along the median from the vertex.
- For a sector of a circle of radius r and angle at the centre 2α radians, G is on the axis of symmetry $\frac{2r\sin\alpha}{3\alpha}$ from the centre.
- For a circular arc of radius r and angle at the centre 2α radians, G is on the axis of symmetry $\frac{r\sin\alpha}{\alpha}$ from the centre.
- For composite bodies you treat each component as a particle at its centre of mass and find the centre of mass of the resulting system of particles. 2.2
- When an object is suspended in equilibrium, its centre of mass is vertically below its point of suspension.
- An object resting on a surface will topple over if a vertical line through its centre of mass does not pass through its region of contact with the surface. 2.3

Links

When designing any vehicle, it is vital that the engineer knows the exact location of its centre of mass.

For a car, the lower the centre of mass, the less the risk that the car will roll over when it is being driven round a corner.

Similarly, for a tractor working on sloping ground, a high centre of mass would be dangerous because the tractor would be likely to topple over.

Revision exercise 1

1 At time $t = 0$ a small package is projected from a point B which is 2.4 m above a point A on horizontal ground. The package is projected with speed 23.75 m s^{-1} at an angle α to the horizontal, where $\tan \alpha = \frac{4}{3}$. The package strikes the ground at the point C, as shown in the diagram. The package is modelled as a particle moving freely under gravity.

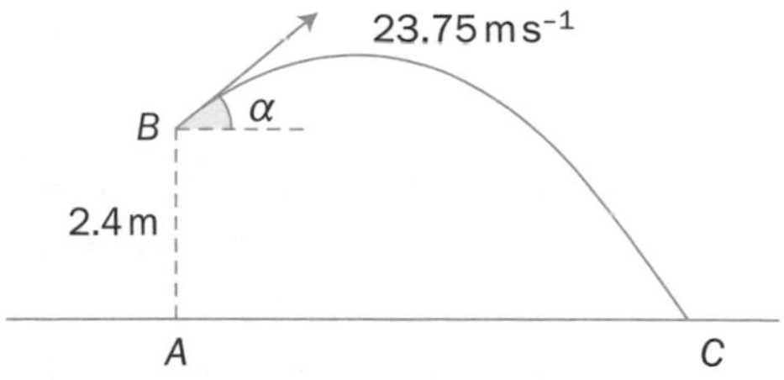

a Find the time taken for the package to reach C.

A lorry moves along the line AC, approaching A with constant speed 18 m s^{-1}. At time $t = 0$ the rear of the lorry passes A and the lorry starts to slow down. It comes to rest T seconds later. The acceleration, a m s^{-2}, of the lorry at time t seconds is given by

$$a = -\frac{1}{4}t^2, \quad 0 \leqslant t \leqslant T$$

b Find the speed of the lorry at time t seconds.

c Hence show that $T = 6$.

d Show that when the package reaches C it is just under 10 m behind the rear of the moving lorry. [(c) Edexcel Limited 2001]

2 A particle P of mass 0.3 kg is moving under the action of a single force, $\mathbf{F}$ newtons. At time t seconds the velocity of P, $\mathbf{v}$ m s^{-1}, is given by

$$\mathbf{v} = 3t^2\mathbf{i} + (6t - 4)\mathbf{j}$$

a Calculate, to 3 significant figures, the magnitude of $\mathbf{F}$ when $t = 2$.

When $t = 0$, P is at the point A. The position vector of A with respect to a fixed origin O is $(3\mathbf{i} - 4\mathbf{j})$ m. When $t = 4$, P is at the point B.

b Find the position vector of B. [(c) Edexcel Limited 2002]

3 A particle P is projected with velocity $(2u\mathbf{i} + 3u\mathbf{j})$ m s^{-1} from a point O on a horizontal plane, where $\mathbf{i}$ and $\mathbf{j}$ are horizontal and vertical unit vectors respectively. The particle P strikes the plane at the point A which is 735 m from O.

a Show that $u = 24.5$.

b Find the time of flight from O to A.

The particle P passes through a point B with speed 65 m s^{-1}.

c Find the height of B above the horizontal plane. [(c) Edexcel Limited 2004]

4

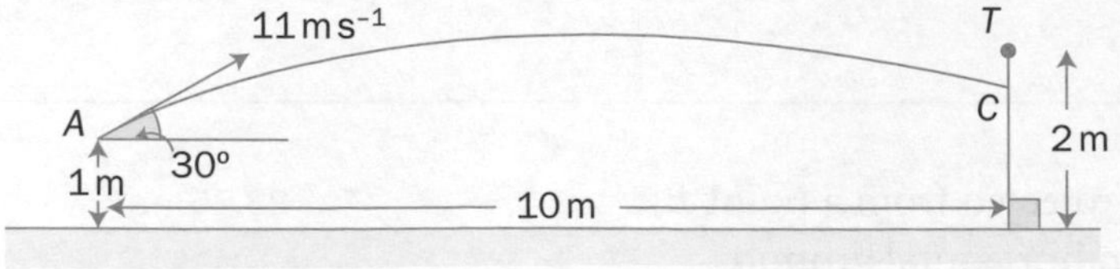

The object of a game is to throw a ball B from a point A to hit a target T which is placed at the top of a vertical pole, as shown in the diagram. The point A is 1 m above horizontal ground and the height of the pole is 2 m. The pole is at a horizontal distance of 10 m from A. The ball B is projected from A with a speed of 11 m s^{-1} at an angle of elevation of 30°. The ball hits the pole at the point C. The ball B and the target T are modelled as particles.

a Calculate, to 2 decimal places, the time taken for B to move from A to C.

b Show that C is approximately 0.63 m below T.

The ball is thrown again from A. The speed of projection of B is increased to V m s^{-1}, the angle of elevation remaining 30°. This time B hits T.

c Calculate the value of V.

d Explain why, in practice, a range of values of V would result in B hitting the target. [(c) Edexcel Limited 2006]

5 A particle P moves on the x-axis. At time t seconds, its acceleration is $(5 - 2t)$ m s^{-2}, measured in the direction of x increasing. When $t = 0$, its velocity is 6 m s^{-1} measured in the direction of x increasing. Find the time when P is instantaneously at rest in the subsequent motion. [(c) Edexcel Limited 2006]

6 A vertical cliff is 73.5 m high. Two stones A and B are projected simultaneously. Stone A is projected horizontally from the top of the cliff with speed 28 m s^{-1}. Stone B is projected from the bottom of the cliff with speed 35 m s^{-1} at an angle α above the horizontal. The stones move freely under gravity in the same vertical plane and collide in mid-air. By considering the horizontal motion of each stone

a prove that $\cos\alpha = \frac{4}{5}$

b find the time which elapses between the instant when the stones are projected and the instant when they collide. [(c) Edexcel Limited 2006]

M2

7 The diagram shows a decoration which is made by cutting two circular discs from a sheet of uniform card. The discs are joined so that they touch at a point D on the circumference of both discs. The discs are coplanar and have centres A and B with radii 10 cm and 20 cm respectively.

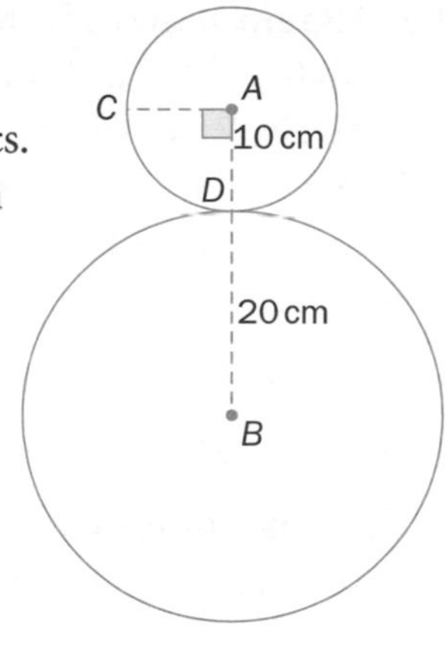

a Find the distance of the centre of mass of the decoration from B.

The point C lies on the circumference of the smaller disc and $C\hat{A}B$ is a right angle. The decoration is freely suspended from C and hangs in equilibrium.

b Find, in degrees to one decimal place, the angle between AB and the vertical.

[(c) Edexcel Limited 2001]

8 The diagram shows a template made by removing a square $WXYZ$ from a uniform triangular lamina ABC. The lamina is isosceles with $CA = CB$ and $AB = 12a$. The midpoint of AB is N and $NC = 8a$. The centre O of the square lies on NC and $ON = 2a$. The sides WX and ZY are parallel to AB and $WZ = 2a$. The centre of mass of the template is at G.

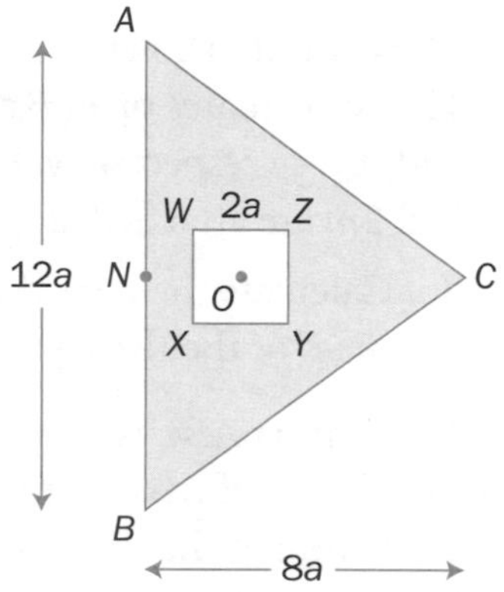

a Show that $NG = \frac{30}{11}a$.

The template has mass M. A small metal stud of mass kM is attached to the template at C. The centre of mass of the combined template and stud lies on YZ. By modelling the stud as a particle

b calculate the value of k.

[(c) Edexcel Limited 2002]

9 A uniform lamina L is formed by taking a uniform square sheet of material $ABCD$, of side 10 cm, and removing the semicircle with diameter AB from the square, as shown in the diagram.

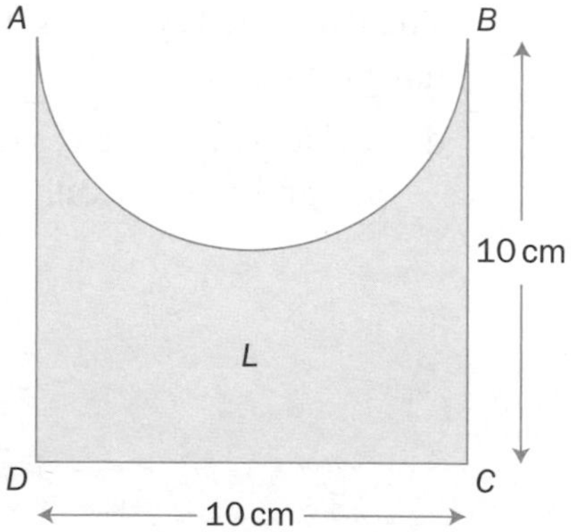

a Find, in cm to 2 decimal places, the distance of the centre of mass of the lamina L from the midpoint of AB.
[The centre of mass of a uniform semicircular lamina, radius a, is at a distance $\frac{4a}{3\pi}$ from the centre of the bounding diameter.]

The lamina is freely suspended from D and hangs at rest.

b Find, in degrees to one decimal place, the angle between CD and the vertical.

[(c) Edexcel Limited 2002]

M2

10 The diagram shows a rectangular lamina *OABC*. The coordinates of *O*, *A*, *B* and *C* are (0,0), (8,0), (8,5) and (0,5) respectively. Particles of mass km, $5m$ and $3m$ are attached to the lamina at *A*, *B* and *C* respectively.
The x-coordinate of the centre of mass of the three particles *without the lamina* is 6.4.

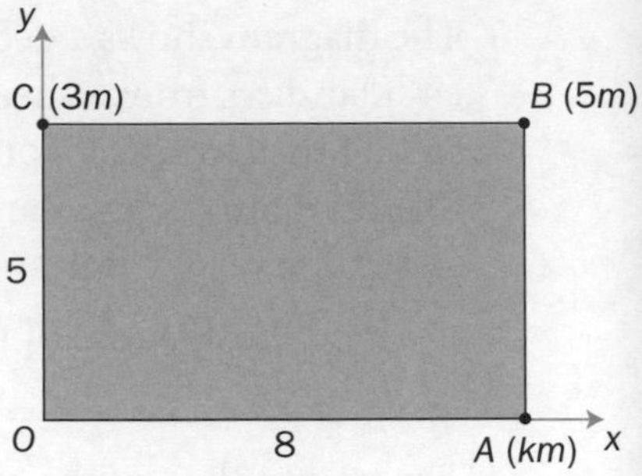

a Show that $k = 7$.

The lamina *OABC* is uniform and has mass $12m$.

b Find the coordinates of the centre of mass of the combined system consisting of the three particles and the lamina.

The combined system is freely suspended from *O* and hangs at rest.

c Find the angle between *OC* and the horizontal.

[(c) Edexcel Limited 2008]

11 The diagram shows a triangular lamina *ABC*. The coordinates of *A*, *B* and *C* are (0,4), (9,0) and (0,–4) respectively. Particles of mass $4m$, $6m$ and $2m$ are attached at *A*, *B* and *C* respectively.

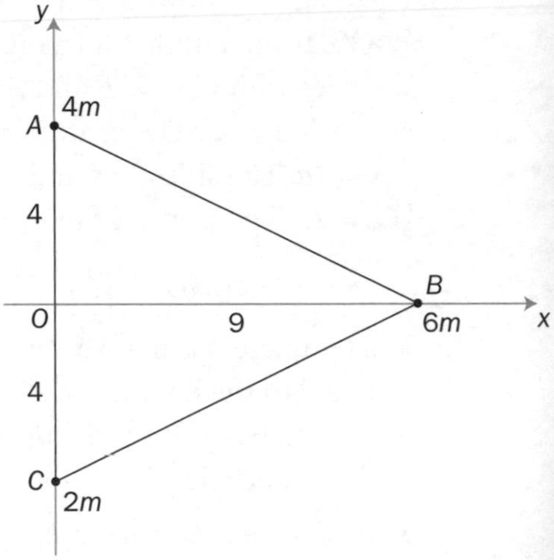

a Calculate the coordinates of the centre of mass of the three particles, *without the lamina*.

The lamina *ABC* is uniform and of mass km. The centre of mass of the combined system consisting of the three particles and the lamina has coordinates $(4, \lambda)$.

b Show that $k = 6$.

c Calculate the value of λ.

The combined system is freely suspended from *O* and hangs at rest.

d Calculate, in degrees to one decimal place, the angle between *AC* and the vertical.

[(c) Edexcel Limited 2006]

M2

3 Work, energy and power

This chapter will show you how to

- calculate the work done by a constant force
- calculate gravitational potential energy and kinetic energy
- use the principle of conservation of mechanical energy
- use the work-energy principle
- calculate power
- relate power and velocity.

Before you start

You should know how to:	Check in:
1 Manipulate algebraic equations.	1 Find v if $\frac{1}{2}mv^2 + mgh = 3mgh$
2 Manipulate vectors.	2 $\mathbf{r} = 5\mathbf{i} + 7\mathbf{j}$, $\mathbf{s} = 3\mathbf{i} - 2\mathbf{j}$ Find **a** $5\mathbf{r} - 4\mathbf{s}$ **b** $\lvert\mathbf{r} + \mathbf{s}\rvert$
3 Recall the equations of motion with constant acceleration.	3 State the equations of motion with constant acceleration.

M2

3.1 Work

A force does work if it alters the motion or position of an object. The amount of work depends on the magnitude of the force and the distance through which its point of application moves.

Gavin and Tracy use pulleys to lift a 50 kg object through 0.5 m.

Gavin uses this arrangement.

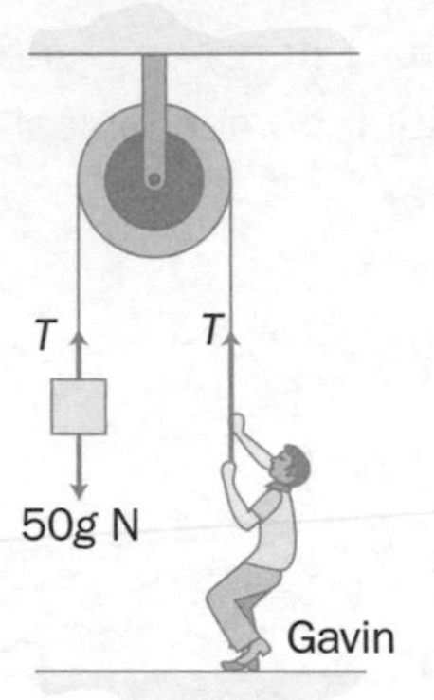

Assume there is no friction and the object rises at constant speed. The mass of Gavin must be more than 50 kg.

Resolve vertically for the object:

$T - 50g = 0$ (no acceleration)

So the tension throughout the rope is $50g$N.

Gavin has to exert a force of $50g$N on the rope and move it down 0.5 m.

Tracy is not as strong as Gavin, but more ingenious.
She uses this arrangement.

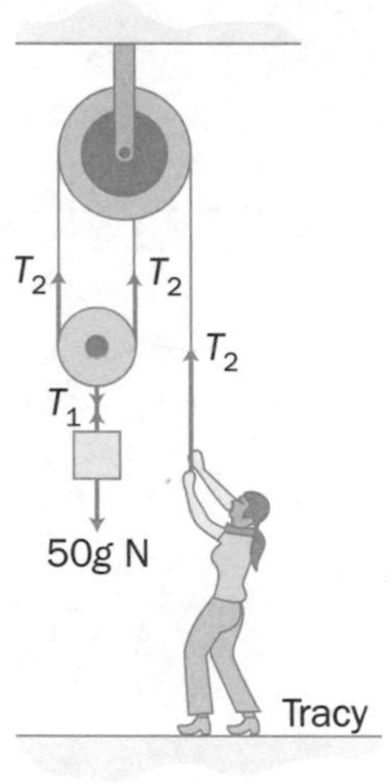

Assume the extra pulley has negligible mass.

Resolve vertically for the object:

$T_1 - 50g = 0$ so $T_1 = 50g$N

Resolve vertically for the small pulley:

$2T_2 - T_1 = 0$ so $T_2 = 25g$N

Tracy has to exert a force of $25g$N. But the rope is 'shared' between the two sides of the small pulley, so to raise the object by 0.5 m, Tracy must pull 1 m of rope through the pulley.

Both Gavin and Tracy do the same amount of work – they both raise 50 kg through a distance of 0.5 m. Tracy exerts half the force that Gavin does, but she exerts it through twice the distance to achieve the same effect.

This leads to the definition of work.

If the point of application of a force F undergoes a displacement s in the direction of the force, then

work done by the force $= F \times s$

The SI **unit of work** is the **joule (J)**.

1 joule is the work done when a force of 1 N is displaced through 1 m.

In the example Gavin does $50g \times 0.5 = 245$ J of work.
Tracy does $25g \times 1 = 245$ J of work.

What matters is the displacement *in the direction of the force.* In the diagram a block is dragged along the ground by a rope. Tension T does work and work is also done against friction F, but reaction R and weight W are perpendicular to the direction of motion and so do no work.

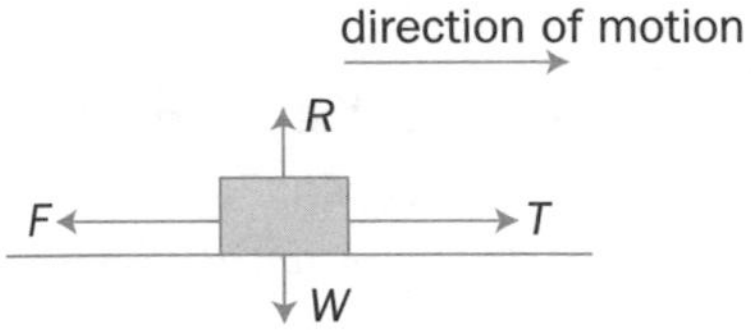

EXAMPLE 1

A block of mass 60 kg is dragged 3 m at constant speed across a horizontal rough plane by a horizontal rope. The coefficient of friction between the block and the plane is 0.4. Find the work done by each of the forces acting on the block.

Resolve vertically:

$$R - 60g = 0$$
$$R = 60g\text{N}$$

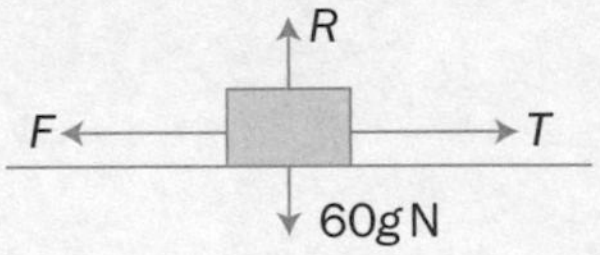

As $\mu = 0.4$, you have

$$F = 0.4 \times R = 24g\text{N}$$

$F = \mu R$

Resolve horizontally:

$$T - F = 0 \quad \text{(no acceleration)}$$
$$T = 24g\text{N}$$

There is no vertical displacement, so the work done by the weight and by R is zero.

Displacement in the direction of T is 3 m, so you have
work done by the tension = $(24g \times 3)$ J = $72g$J

Displacement in the direction of F is −3 m so you have
work done by the friction = $(24g \times (-3))$ J = $-72g$J

Work related to friction has a negative value. You say either that $-72g$J of work is done **by** friction or that $72g$J of work is done **against** friction (work done by the system to overcome the friction).

When an object is lifted, work is done against gravity.

Suppose an object of mass m is raised at constant speed by a force T.

Resolve vertically:

$$T - mg = 0$$
$$T = mg$$

If the object is raised through a distance h
work done by $T = mgh$
work done by the weight $= -mgh$

Even when the object does not travel vertically, if its final position is h above its initial position the work against gravity is mgh.

M2

EXAMPLE 2

A plank of length 5 m is inclined so that the higher end is 3 m above the lower. A block of mass 40 kg is towed at constant speed up the plank against a friction force of 120 N. Find the total work done by the towing force.

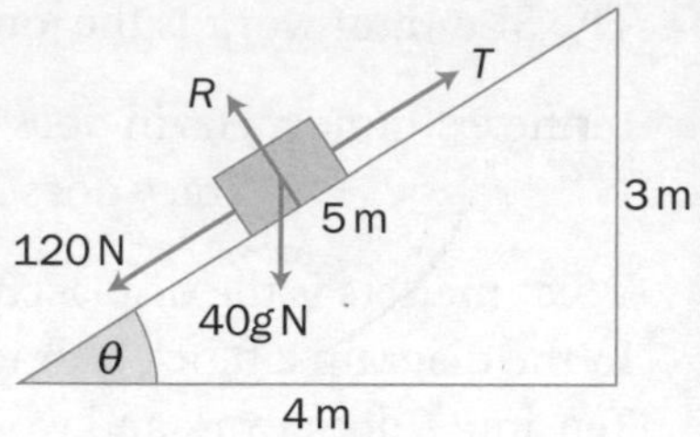

The block is displaced –5 m in the direction of the friction force, so

work done *against* friction = (120 × 5) J = 600 J

The block is raised through a height of 3 m, so

work done *against* gravity = ($40g \times 3$) J = 1176 J

So the total work done in raising the block

= (600 + 1176) J = 1776 J

There is an alternative approach to this problem:

Resolve parallel to the slope:

$T - 120 - 40g\sin\theta = 0$

You know $\sin\theta = \frac{3}{5}$, so

$$T = 120 + 24g = 355.2$$

giving a towing force of 355.2 N

The block is displaced 5 m in the direction of T, so

work done = (355.2 × 5) J = 1776 J

The concept of work done against gravity is particularly useful if the path of the object is not a straight line.

EXAMPLE 3

An 8 kg block is pulled at constant speed up a surface forming a quarter of a circle, radius 2 m, against a constant friction of 45 N. Find the total work done.

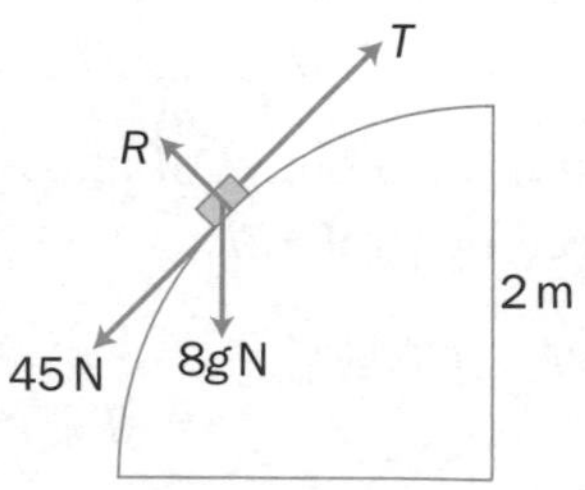

The object is raised through 2 m, so

work done against gravity = ($8g \times 2$) J = 156.8 J

The object travels a distance of $-\left(\frac{1}{4} \times 4\pi\right)$ m = $-\pi$ m in the direction of the friction force, so

work done against friction = 45π J = 141.4 J

Hence, total work done = (156.8 + 141.4) J = 298.2 J

M2

You will meet situations where there is **displacement at an angle to the force**. There are two ways of dealing with this, each leading to the same result.

For example, if a block is dragged along a horizontal surface using a rope, the rope may be inclined to the horizontal.

Method 1

In the diagram the object is displaced along AB, a distance s at an angle θ to the direction of the applied force F N.

The point of application of F moves from A to B, but the displacement in the direction of F is AC, where

$$AC = s\cos\theta$$

This gives work done by $F = Fs\cos\theta$

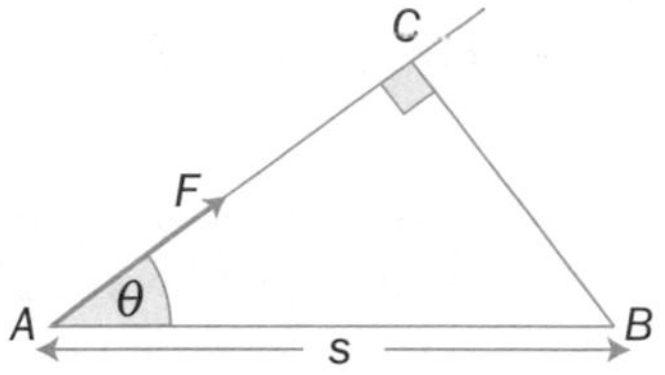

Method 2

The force F can be resolved into two components, parallel and perpendicular to the direction of the displacement, as shown.

The perpendicular component, $F\sin\theta$, does no work because there is no displacement in that direction.

The parallel component, $F\cos\theta$, is displaced a distance s.
So

$$\text{work done by } F = (F\cos\theta) \times s = Fs\cos\theta$$

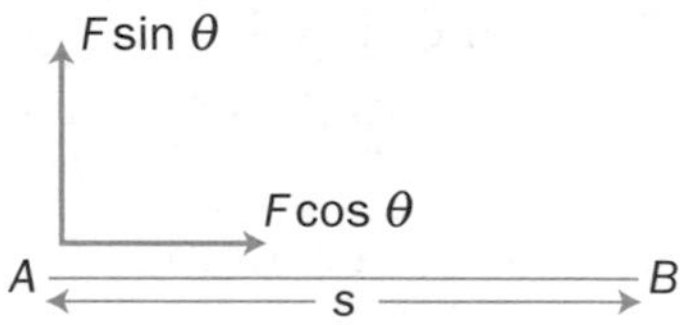

If the point of application of a force F undergoes a displacement s at an angle θ to the direction of F, then
work done by $F = Fs\cos\theta$

M2

EXAMPLE 4

A packing case is dragged 8 m along a horizontal surface by a rope inclined at 40° to the horizontal. The tension in the rope is 500 N. Find the work done by the tension.

The horizontal component of the tension is $500\cos 40°$.
Displacement is 8 m horizontally.

Work done $= (500\cos 40° \times 8)$ J
$= 3060$ J to 3 s.f.

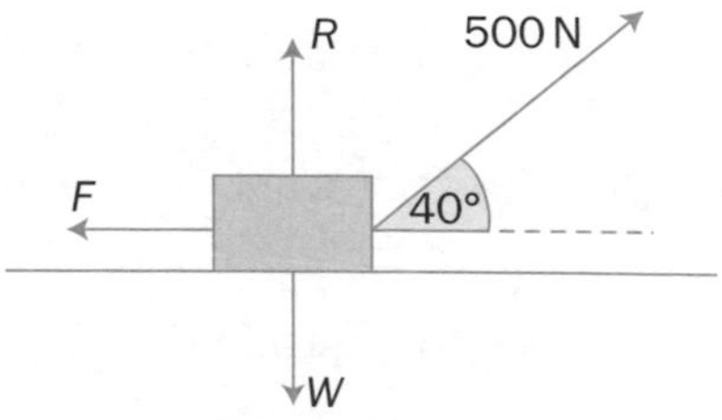

Exercise 3.1

1 A force of 300 N is applied to a particle, which moves 12 m across a horizontal surface. Find the work done by the force if it is

a horizontal b inclined at 35° to the horizontal.

2 A car is travelling at a constant speed of 120 km h^{-1} against a constant resistance force of 150 N. Find the work done by the engine in 1 second.

3 Find the work done by a crane which raises a load of 250 kg at a constant speed through a distance of 5.6 m.

4 A man of mass 85 kg needs to move a load of mass 30 kg from ground level on to scaffolding 6 m up. How much work will he do if

a he hoists it up using a smooth pulley and a light rope?

b he carries it up a ladder?

5 A winch is able to do 1400 J of work in each second. It takes 49 s to raise a 200 kg load from the ground to the top of a building. Find the height of the building.

6 Blocks, each of mass 5 kg and height 10 cm, are lying side by side on the ground. How much work would be involved in making a stack

a ten blocks high b n blocks high?

7 A block of mass 15 kg is pulled at constant speed for a distance of 12 m across a rough horizontal plane. The coefficient of friction between the plane and the block is 0.6. Find the work done against friction.

8 A horizontal force is applied to a 6 kg body so that it accelerates uniformly from rest and moves across a horizontal plane against a constant frictional resistance of 30 N. After it has travelled 16 m, it has a speed of 4 m s^{-1}. Find

a the acceleration of the body

b the magnitude of applied force

c the work done by the applied force.

9 A winch raises an object of mass 20 kg from rest with an acceleration of 0.2 m s^{-2}. Find how much work the winch does in the first 12 seconds.

10 Find the work done in pulling a packing case of mass 80 kg at constant speed for a distance of 15 m against a constant resistance of 150 N

a on a horizontal surface

b up a slope inclined at $\arcsin \frac{1}{8}$ to the horizontal.

M2

11 A block of mass 10 kg is pulled at constant speed a distance of 8 m up a slope inclined at 20° to the horizontal. Find the work done if

a the surface is smooth

b the coefficient of friction between the block and the slope is 0.4.

12 A block of mass 5 kg is at rest on a rough horizontal surface with coefficient of friction μ. A horizontal force of 25 N is applied to the block. This force does 1000 J of work in the next 10 s.

a Show that $\mu = \frac{3}{7}$.

b Find how much work the force would have done in the 10-second period if it had been applied in a direction 30° above the horizontal.

13 A ramp joining a point A to a point C, which is 8 m higher than A, consists of a straight section and an arc of a circle centre B, as shown. A trolley, which has mass 10 kg and a constant resistance to motion of 140 N, is pulled at a steady speed from A to C. Find the total work done.

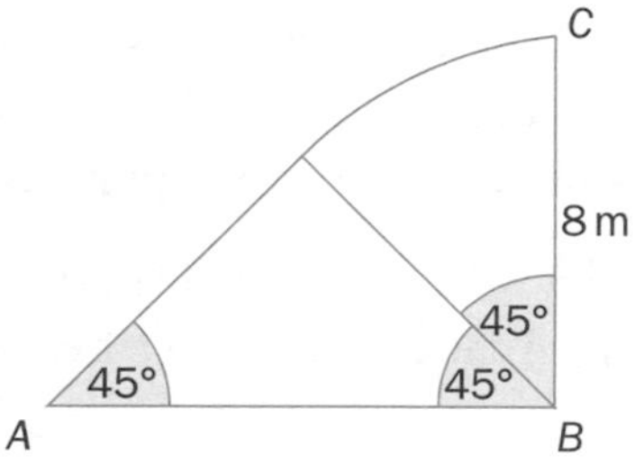

M2

14 A body of mass 10 kg is at rest on a rough horizontal surface. The coefficient of friction between the body and the surface is 0.5. A force of 100 N is applied to the body for a period of 20 seconds in one of three different directions, as shown below.

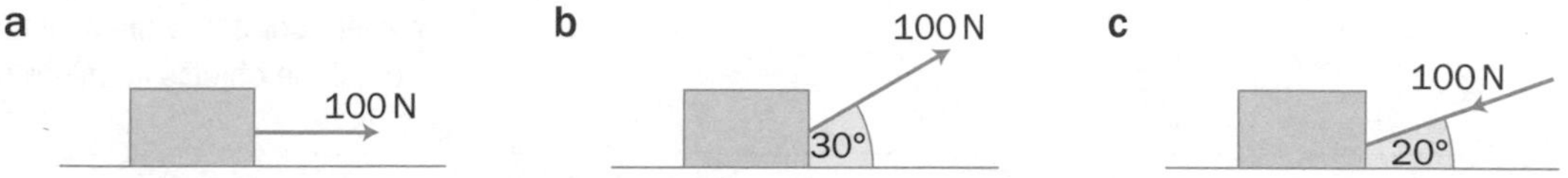

In each case, find the distance travelled by the body and hence the work done by the applied force.

15 A rough surface is inclined to the horizontal at 25°. A body of mass 20 kg is pulled for 15 m up the line of greatest slope. The coefficient of friction between the body and the surface is 0.25. Find the total amount of work done on the body if

a the body travels all the way at constant speed

b the body starts from rest and takes 6 s to travel the 15 m with constant acceleration.

16 A rough slope, of length x, is inclined at an angle α to the horizontal. A block of mass m is towed up the slope at constant speed. The coefficient of friction between the block and the slope is μ.

a Show that the total amount of work done on the block is given by $mgx(\sin\alpha + \mu\cos\alpha)$

The block is now towed back down the slope. The same amount of work is done as on the upward journey. The block starts from rest at the top of the slope.

b Find the speed at which the block is travelling when it reaches the bottom.

3.2 Energy

Suppose you raise an object of mass 20 kg by 10 m.
The work done against gravity is $(20g \times 10)$ J = 1960 J.

If you let the object sink back to its original level, the force due to gravity will do 1960 J of work.

By raising the object you *store* 1960 J of work, which you can *retrieve* by lowering the object.

Stored work is called **energy**. The unit of energy is the joule.

This way of storing work is used in many ways. For example, some wall-clocks are powered by weights on a chain passing over a cog. You start the heavier weight at the top. As it gradually descends, it does the work needed to drive the clock.

Energy of position is **gravitational potential energy (GPE)**. It depends on the mass of the object and the height of the object relative to a chosen zero level. In a given problem you set a zero level and measure GPE relative to that.

The work done against gravity in raising an object of mass m kg from the zero level to a height of h metres is mgh J.

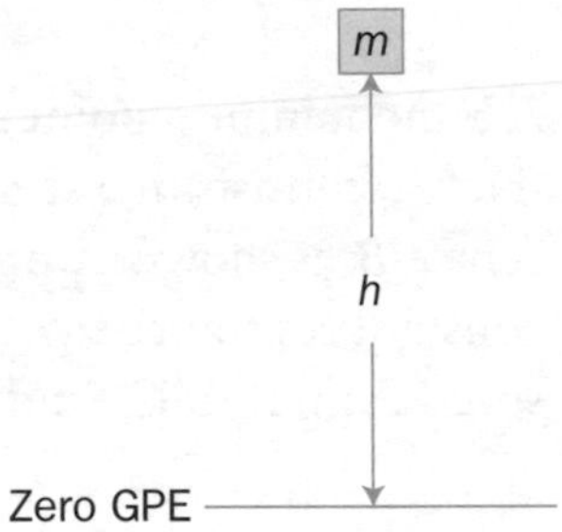

For an object of mass m kg at a height h m from the zero level
Gravitational potential energy = mgh J

For an object below the zero level, GPE is negative. This is not a problem, as change in energy is what matters. You can avoid negative GPE by setting the zero level at or below the minimum height of all the objects in the problem.

EXAMPLE 1

A crane carries a load of 400 kg. It raises it from ground level to a height of 30 m, then lowers it on to a platform 12 m above the ground. Find the change in potential energy in each stage.

Take ground level to be zero GPE.
The diagram shows the initial (A), intermediate (B) and final (C) positions of the load.

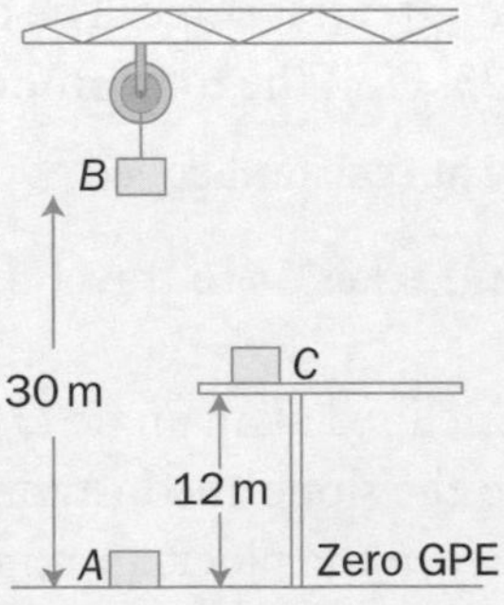

If no diagram is provided in the question, your own sketch will help you to visualise the problem.

At A: GPE = 0 J
At B: GPE = $(400g \times 30)$ J = 117 600 J
At C: GPE = $(400g \times 12)$ J = 47 040 J

During stage 1, GPE increases by 117 600 J.

During stage 2, GPE decreases by $(117\,600 - 47\,040)$ J = 70 560 J

You also store work when you accelerate an object. For example, suppose you apply a force of 100 N to an object of mass 20 kg, initially at rest, while it travels 10 m.

The work done is $(100 \times 10)\,\text{J} = 1000\,\text{J}$.
The object accelerates to $10\,\text{m}\,\text{s}^{-1}$.

From $F = ma$, you get $a = 5\,\text{m}\,\text{s}^{-2}$
From $v^2 = u^2 + 2as$, you get $v = 10\,\text{m}\,\text{s}^{-1}$

You would need to apply another force to stop the object.
Work would be done against that force.
For example, if you apply a frictional force of 200 N, the object slows to rest in 5 m.
The work done against friction is $(200 \times 5)\,\text{J} = 1000\,\text{J}$.

From $F = ma$, you get $a = -10\,\text{m}\,\text{s}^{-2}$
From $v^2 = u^2 + 2as$, you get $s = 5\,\text{m}$

By giving the object a speed of $10\,\text{m}\,\text{s}^{-1}$, you stored 1000 J of work, which was retrieved when the object was brought to rest.

The work capacity of an object due to its motion is called kinetic energy (KE). It depends on the mass of the object and its velocity. You can derive the formula as follows.

Suppose you apply a force F N to a stationary object of mass m kg for a distance s m. It accelerates at $a\,\text{m}\,\text{s}^{-2}$ to a velocity $v\,\text{m}\,\text{s}^{-1}$.

$$v^2 = 2as$$

$$as = \frac{1}{2}v^2 \qquad [1]$$

$$\text{work done} = Fs$$

$$= mas \qquad [2]$$

Substitute from [1] into [2]:

$$\text{work done} = \frac{1}{2}mv^2\,\text{J}$$

This work corresponds to the kinetic energy of the object.

An object of mass m kg travelling at $v\,\text{m}\,\text{s}^{-1}$ has

$$\text{Kinetic energy} = \frac{1}{2}mv^2\,\text{J}$$

M2

EXAMPLE 2

A particle of mass 4 kg is initially at rest. It is acted upon by a force of 60 N for a period of 8 seconds. Find its kinetic energy at the end of this time.

Use $F = ma$:

$$60 = 4a \quad \text{and so} \quad a = 15\,\text{m}\,\text{s}^{-2}$$

Use $v = u + at$, where $a = 15\,\text{m}\,\text{s}^{-2}$, $t = 8\,\text{s}$ and $u = 0$:

$$v = (0 + 15 \times 8)\,\text{m}\,\text{s}^{-1} = 120\,\text{m}\,\text{s}^{-1}$$

Hence kinetic energy $= \left(\frac{1}{2} \times 4 \times 120^2\right)\text{J} = 28\,800\,\text{J}$ or $28.8\,\text{kJ}$

Alternatively, use $s = ut + \frac{1}{2}at^2$:

$$s = \left(\frac{1}{2} \times 15 \times 8^2\right)\text{m} = 480\,\text{m}$$

Therefore, you have

work done by force $= (60 \times 480)\,\text{J} = 28\,800\,\text{J}$

As all of this work went into accelerating the particle, the kinetic energy is 28 800 J.

M2

KE and GPE are forms of mechanical energy.
There is a third form of mechanical energy – elastic potential energy – but this is not covered in this module.

There are non-mechanical forms of energy, such as heat, light and electrical energy. These are often generated mechanically, such as when objects become hot because of friction.

The mechanical energy of a system is changed if an external force does work on it. However, in some situations external forces are negligible, and so no work is done on the system.

Suppose, for example, that you suspend a particle by a light string and set it swinging as a pendulum. Assume there is no air resistance. Consider the stages of motion of the particle:

On the downswing, the height of the particle and therefore its GPE decrease, but its speed and therefore its KE increase.

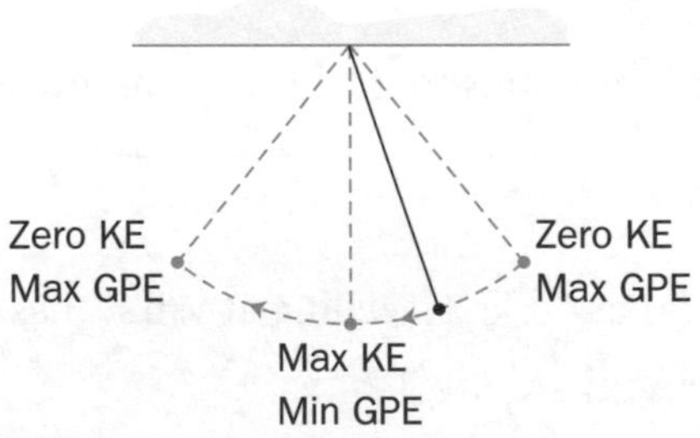

When the particle reaches its lowest position, its GPE is at a minimum and its KE is at a maximum.

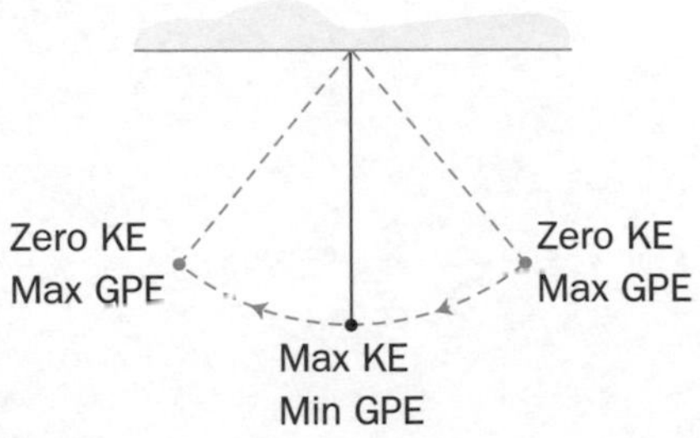

As the particle swings up from its lowest point, work is done against gravity. The particle gains height (GPE increases) and slows down (its KE decreases).

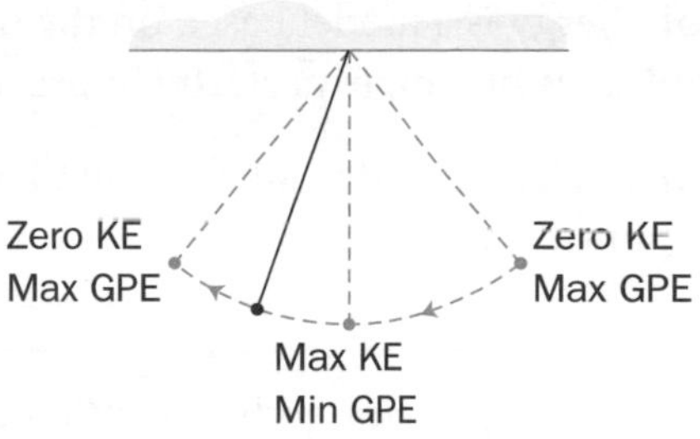

The particle stops, at which point it has zero KE and maximum GPE.

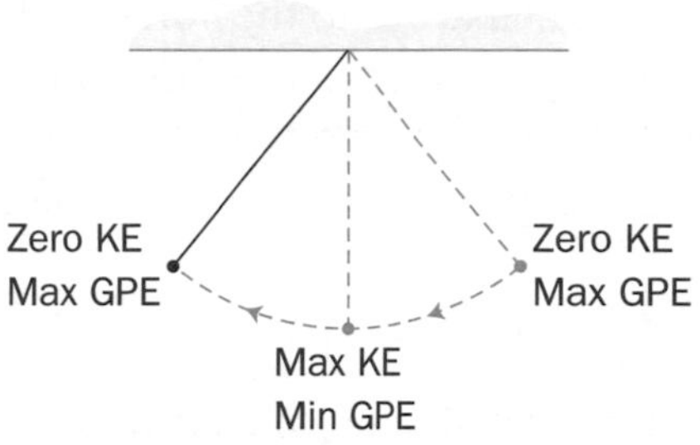

The particle then swings down again and the pattern starts again.

In this idealised model, the sequence repeats forever. On the way up, KE is converted into GPE. On the way down, GPE is converted into KE. The total energy of the particle is exactly the same at all points.

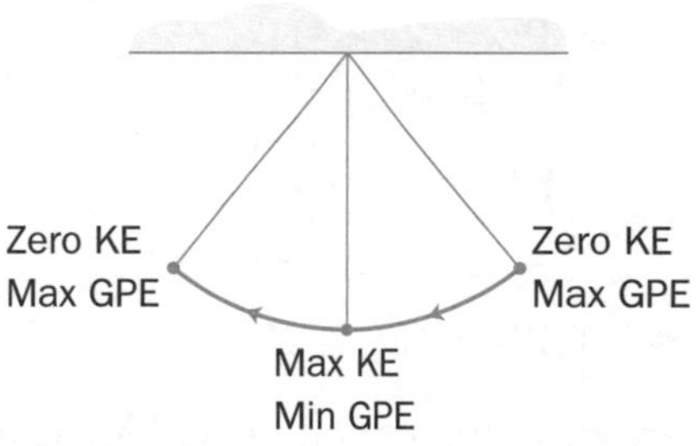

M2

The total mechanical energy of a system only changes because of one of the following:

- An external force, other than gravity, acts on the system in such a way that work is done.
- There are sudden changes in the motion of the system. This happens if parts of the system collide, or if strings connecting parts of the system are jerked taut.

If the external force does work on the system, its energy increases. If the system does work against the external force (e.g. against friction), its energy decreases.

Although the forces involved in such sudden changes are internal to the system, there is usually a loss of energy in the form of heat or sound.

This is the **principle of conservation of mechanical energy**.

> The total mechanical energy of a system remains constant provided no external work is done and there are no sudden changes in the motion of the system.

Conservation of energy is a model based on the usual assumptions and does not apply perfectly in practice.

However, external forces are often small enough for the principle to have useful applications.

EXAMPLE 3

A particle of mass 2 kg is released from rest and slides down a smooth plane inclined at 30° to the horizontal. Find the speed of the particle after it has travelled 8 m.

The start and finish positions of the particle are *A* and *B* in the diagram.
The speed of the particle at *B* is *v*.

A is at a height of $8 \sin 30° \text{ m} = 4 \text{ m}$ above *B*.

You have the following energy situations:

At *A*: $\text{KE} = 0 \text{ J}$

$\text{GPE} = (2g \times 4) \text{ J}$

$= 78.4 \text{ J}$

At *B*: $\text{KE} = \frac{1}{2} \times 2 \times v^2$

$= v^2$

$\text{GPE} = 0 \text{ J}$

There are no external forces and no sudden changes, so energy is conserved:

Total energy at *B* = Total energy at *A*
which gives $v^2 = 78.4$
so the velocity *v* is 8.85 m s^{-1}

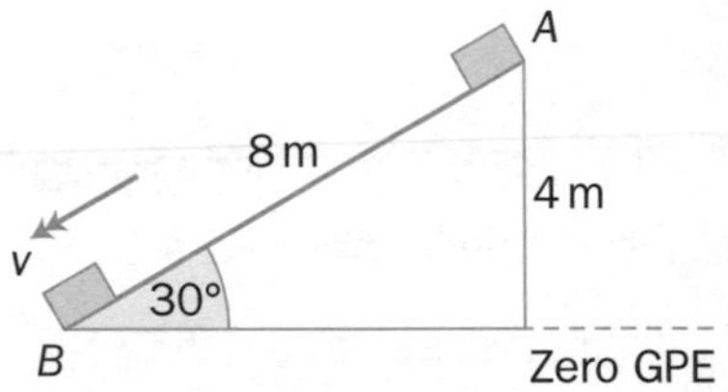

You could have solved this problem using the constant acceleration equations. However, using energy has one major advantage: there is no need to assume that *AB* is a straight line. If the particle follows a curved path from *A* to *B*, so that acceleration is no longer constant, the energy approach shows that its final speed will still be 8.85 m s^{-1}.

If external forces act on a system so that work is done, mechanical energy is not conserved. However, you can still make use of energy to solve problems provided you know the total amount of work done on the system.

The **work-energy principle**
The total work done on a system equals the change in mechanical energy of the system.

The following examples introduce a resistance force.

EXAMPLE 4

Particles P and Q, of mass 6 kg and 2 kg respectively, are connected by a light string of length 8 m passing over a smooth pulley at the top of a double 30° slope, as shown. The coefficient of friction for both particles is 0.1. The system starts from rest with the string taut and P at the top of the slope. Find the velocity of the particles when Q reaches the top of the slope.

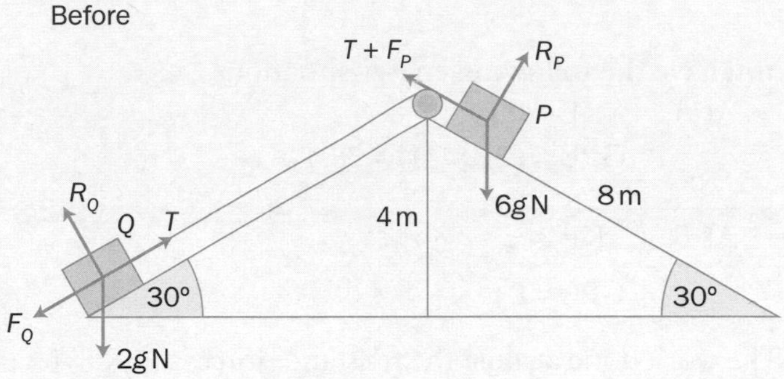

Resolve perpendicular to the slope for P:

$R_P - 6g\cos 30° = 0$ and so $R_P = 50.9\,\text{N}$

This gives $F_P = 5.09\,\text{N}$

Resolve perpendicular to the slope for Q:

$R_Q - 2g\cos 30° = 0$ and so $R_Q = 17.0\,\text{N}$

This gives $F_Q = 1.70\,\text{N}$

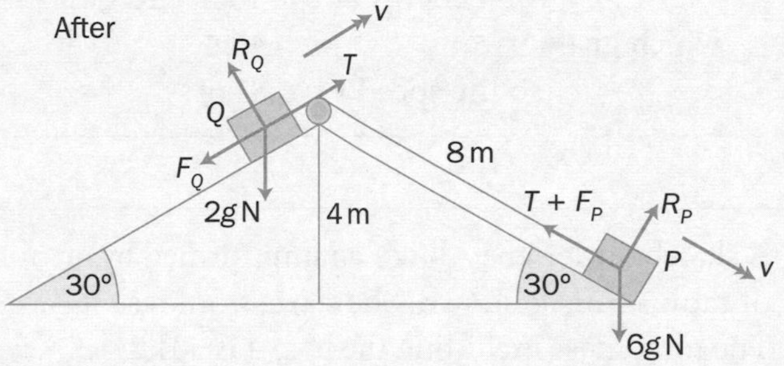

The particles move −8 m in the direction of their friction forces, so:

work done against friction $= (1.70 \times 8 + 5.09 \times 8)\,\text{J}$

$= 54.3\,\text{J}$

Take the bottom of the slopes as the zero GPE level.

Initially the only energy is the GPE of P

initial energy $= (6g \times 4)\,\text{J}$

$= 24g\,\text{J}$

The final energy of the system is the GPE of Q together with the KE of both particles

final energy $= (2g \times 4 + v^2 + 3v^2)\,\text{J}$

$= (8g + 4v^2)\,\text{J}$

Use the work–energy principle:

final energy = initial energy − work done

$8g + 4v^2 = 24g - 54.3$

This gives $v^2 = 25.6$

and so the velocity v is $5.06\,\text{m s}^{-1}$

EXAMPLE 5

A particle of mass 2 kg is released from rest and slides down a plane inclined at 30° to the horizontal. There is a constant resistance force of 4 N. Find the speed of the particle after it has travelled 8 metres.

This is the same example given in Example 3 but with a resistance force.

Sketch a diagram.

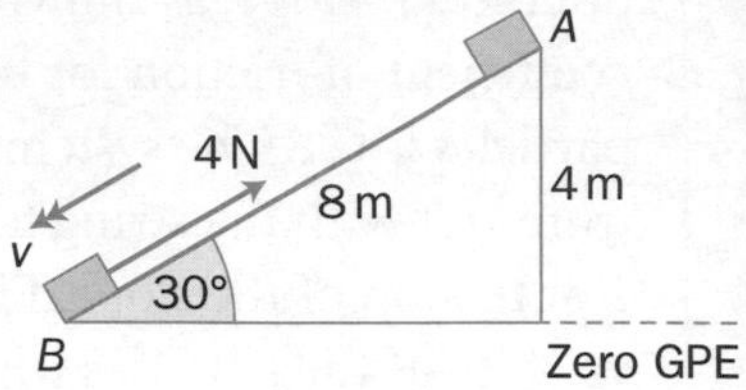

You have the following energy situations:

At A: KE = 0 J

GPE = $(2g \times 4)$ J = 78.4 J

At B: KE = $\frac{1}{2} \times 2 \times v^2 = v^2$

GPE = 0 J

The work done against the resistance force is (4×8) J = 32 J, so

Change of energy of system = −32 J

Total energy at B = Total energy at A − 32

which gives $v^2 = 46.4$

so the speed v is 6.81 m s^{-1}

EXAMPLE 6

A skateboarder goes down a ramp formed by an arc of a circle of radius 5 m, as shown. She starts from rest at A.
The total mass including the board is 50 kg.
Find the speed, v m s^{-1}, with which she leaves the ramp at B if

a friction is negligible

b there is a constant resistance force of 10 N.

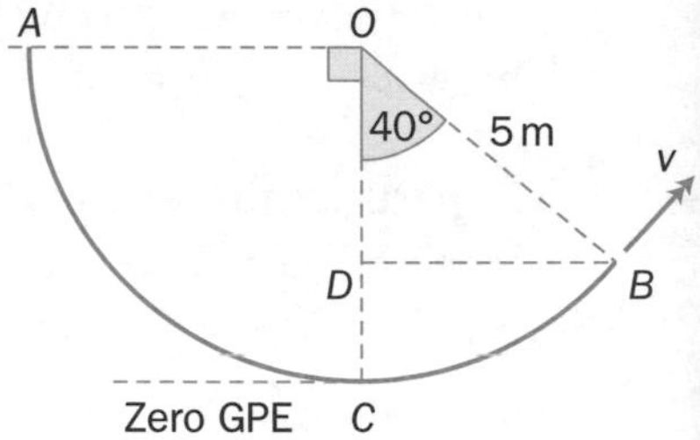

Take C, the bottom of the curve, to be the zero GPE level.

The height of B is $CD = (5 - 5\cos 40°)$ m

= 1.17 m

a You have the following energy situations:

At A: KE = 0 J

GPE = $(50g \times 5)$ J

= 2450 J

At B: KE = $\frac{1}{2} \times 50 \times v^2$

= $25v^2$

GPE = $(50g \times 1.17)$ J

= 573.2 J

Friction is negligible, so energy is conserved.

You have total energy at B = total energy at A

which gives $25v^2 + 573.2 = 2450$

so the speed v is 8.66 m s^{-1}

EXAMPLE 6 (CONT.)

b Work is done against a constant resistance force of 10 N. The distance moved against this force is the arc length of the ramp.

$$\text{Arc length} = \left(2\pi \times 5 \times \frac{130}{360}\right)\text{m} = 11.34\,\text{m}$$

Work done against the resistance = 10 × 11.34 = 113.4 J

This is the change in energy of the system.
You have total energy at B = total energy at A – 113.4
which gives $25v^2 + 573.2 = 2450 - 113.4$
so the speed v is 8.40 m s^{-1}

EXAMPLE 7

Particles A and B, of mass 2 kg and 5 kg respectively, are connected by a light inextensible string passing over a light smooth pulley. The particles are held level and at rest, then released. Find their speed v when they reach 3 m apart.

Take the start position of the two particles as the zero GPE level.

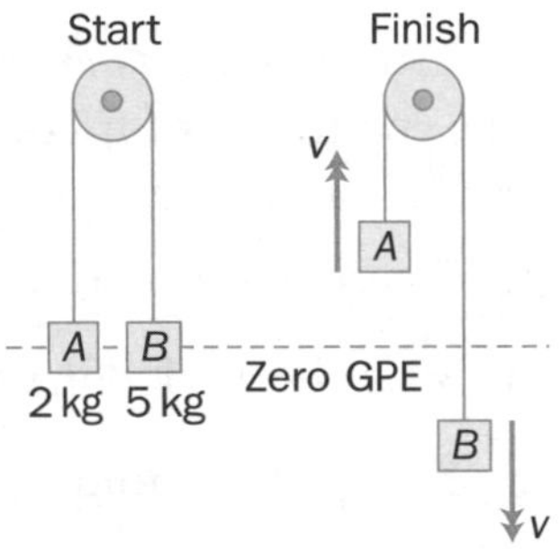

A is 1.5 m above the starting level and *B* is 1.5 m below.

You have the following energy situations:

At start: KE = 0 J
GPE = 0 J

At finish: $\text{KE} = \frac{1}{2} \times 2 \times v^2 + \frac{1}{2} \times 5 \times v^2$
$= 3.5v^2$

$\text{GPE} = (2g \times 1.5 + 5g \times (-1.5))\,\text{J}$
$= -44.1\,\text{J}$

There is no external force so energy is conserved. You have
total energy at finish = total energy at start
which gives $3.5v^2 - 44.1 = 0$
so the speed v is 3.55 m s^{-1}

Exercise 3.2

1 A body of mass 3 kg is dropped from the top of a tower of height h m. The body starts at rest and reaches the bottom of the tower travelling at 28 m s^{-1}. Find

a an expression for the potential energy of the body at the start, taking the bottom of the tower as the zero level

b the kinetic energy of the body when it reaches the bottom

c the value of h.

M2

2 A particle of mass 2 kg is travelling on a rough horizontal surface against a constant resistance force of 0.5 N.
The particle passes through a point *P* with speed $8\,\text{m s}^{-1}$ and later through a point *Q* with speed $4\,\text{m s}^{-1}$.
Find

a the kinetic energy of the particle at *P*

b the kinetic energy of the particle at *Q*

c the work done against the resistance force in travelling from *P* to *Q*

d the distance *PQ*.

3 A ball of mass 0.4 kg is thrown vertically into the air at a speed of $25\,\text{m s}^{-1}$. Assuming that air resistance is negligible, use energy methods to find the speed at which the ball is moving when it reaches a height of 20 m. Is the mass of the ball a necessary piece of information?

4 A particle of mass *m* is sliding on a rough horizontal surface. The coefficient of friction is μ. It passes through a point *A* travelling at $6\,\text{m s}^{-1}$ and later through a point *B* at $2\,\text{m s}^{-1}$.

a Find the loss of kinetic energy in travelling from *A* to *B*.

b State the work done against friction during this time.

c If $AB = 10\,\text{m}$, find μ.

5 A child of mass 25 kg moves from rest down a slide.
The total drop in height is 4 metres.

a Assuming friction is negligible, find the speed of the child at the bottom of the slide.

b In fact, the child reaches the bottom travelling at $6\,\text{m s}^{-1}$.
The length of the slide is 6 m.
Find **i** the work done against friction
ii the average friction force.

6 A particle slides from rest down a smooth plane inclined at 30° to the horizontal. It reaches the bottom of the slope travelling at $14\,\text{m s}^{-1}$. Find the length of the slope.

7 A particle of mass 2 kg starts from rest at A on a smooth 20° slope. A force of 120 N is applied parallel to the slope, moving the particle up the slope to B, where $AB = 3$ m. The force then stops acting and the particle continues up the slope, coming instantaneously to rest at C before sliding back down to A. Using energy methods find

a the speed of the particle at B

b the distance BC

c the speed of the particle when it returns to A.

8 A particle slides from rest for 3 m down a roof inclined at 40° to the horizontal, then falls to the ground 6 m below, as shown.

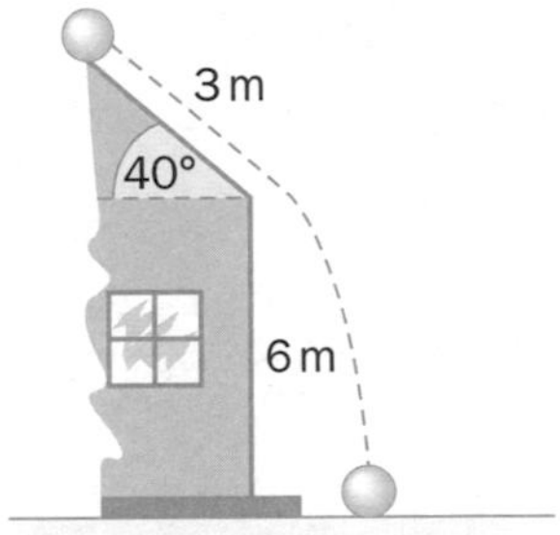

Find the speed with which the particle hits the ground if

a the roof is smooth

b the roof is rough with $\mu = 0.2$.

9 A ball of mass 2 kg falls vertically. It passes a point A with speed 2 m s^{-1} and a point B, 4 m below A, with speed 8 m s^{-1}. Air resistance is assumed to be a constant R N. Find the value of R.

10 Particles A and B, of mass 0.5 kg and 1.5 kg respectively, are connected by a light inextensible string of length 2.5 m. Particle A rests on a smooth horizontal surface at a distance of 1.5 m from its edge. The string passes over the edge of the surface and particle B hangs suspended. The system is held at rest with the string just taut and then released. Use energy methods to find the speed of A when it reaches the edge of the surface. The surface is at least 2.5 m above the ground.

M2

11 Particles A and B, of mass 1 kg and 4 kg respectively, are connected by a light inextensible string of length 4 m. The string passes over a small, light smooth pulley 3 m above the ground. The particles are held at rest with A on the ground and B hanging with the string taut. The system is then released.

a Find the speed of the particles as they pass each other.

b Explain why you could not have used conservation of energy if the system had started with A on the ground and B held level with the pulley.

12 Particles P and Q have masses of 2 kg and 4 kg respectively. They are connected by a light string passing over a smooth pulley which is 5 m above the ground and at the top of a smooth 40° slope, as shown. P starts at the bottom of the slope with the string taut and Q level with the pulley. The system is released from rest.

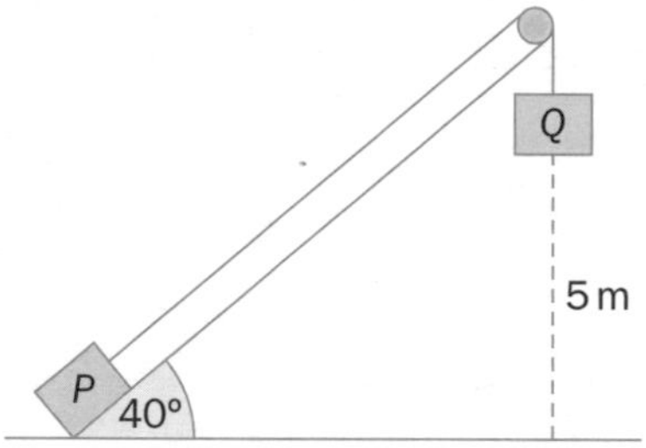

a Find the speed with which Q hits the ground.

Q hits the ground and the string goes slack.

b Find the speed with which P reaches the pulley.

13 A rough roof slopes down at angles of 30° and 60° to the horizontal on either side of its ridge. A particle P, of mass 4 kg, is placed on the 30° slope. A particle Q, of mass 12 kg, is placed on the 60° slope. The particles are connected by a light string passing over a smooth pulley at the ridge. The system is released from rest with the string taut and perpendicular to the ridge. The coefficient of friction for each particle is 0.5. Find the speed with which the particles are travelling after they have moved 4 m (given that they have not reached the pulley or the roof in this time).

14 **a** A particle of mass m is attached to the end A of a light rod OA of length a. O is freely hinged to a fixed point and the rod is held in a horizontal position before being released. Find the speed of the particle when the rod makes an angle θ to the downward vertical.

b If the particle in part **a** had been initially projected downwards with speed u, find the value of u for which the rod would just travel round a complete circle.

15 A light rod AB of length $3a$ has particles of mass m attached at A and at B. O is a point on the rod where $OA = a$. The rod can rotate in a vertical plane about O. The rod is held in a horizontal position and then released. Find the maximum speed of B in the subsequent motion.

16 Particles of mass m and $2m$ are connected together by a light inextensible string. Initially, the particles lie at opposite edges of a smooth horizontal table with the string just taut. One of the particles is then nudged over the edge of the table. Find the ratio between the two possible speeds of the system when the other particle reaches the edge of the table.

17 A particle is projected with velocity V up a rough plane inclined at an angle θ to the horizontal. The coefficient of friction between the particle and the plane is μ, where μ is sufficiently small so that the particle can slide from rest down the plane. The particle starts at a point A, travels up the plane then slides back down through A. At some point B below A on the plane the particle is again travelling with speed V. Show that

$$AB = \frac{\mu V^2 \cos\theta}{g\left(\sin^2\theta - \mu^2\cos^2\theta\right)}$$

M2

3.3 Power

If a person of mass 90 kg climbs a flight of stairs taking them to a height of 15 m, they do $90g \times 15 = 13\,230$ J of work. This is the same whether they run up the stairs or walk up slowly.
However, the effects on their breathing and heart rate vary depending on how quickly they do the work.
This is the **power** they exert.

- Power is the rate at which work is done
- The SI unit of power is the **watt (W)**
- **1 W** is the rate of working of **1 joule per second**

EXAMPLE 1

A crane lifts a load of 50 kg to a height of 12 m at a steady speed of 0.6 m s^{-1}. Find the power required.

The work done against gravity is $(50g \times 12)$ J = 5880 J

Assuming that you can neglect any resistance forces, this is the work done by the crane.
You have time taken to lift load = $(12 \div 0.6)$ s = 20 s
rate of working of crane = $5880 \div 20 = 294$

So the power required is 294 W.

EXAMPLE 2

A car of mass 900 kg moves at a steady speed of 15 m s^{-1} up a slope inclined to the horizontal at 10° against a constant resistance force of 400 N.
Find the power output of the engine.

In 1 s the car goes 15 m up the slope, raising it through a height h where

$h = 15 \sin 10°$ m = 2.60 m

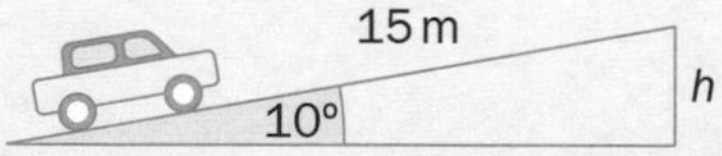

Work done against gravity = $(900g \times 2.60)$ J = 22 974 J
Work done against resistance = (400×15) J = 6000 J

So total work done in each second = 28 974 J

Hence, to 3 s.f., the rate of working = 29 000 W or 29.0 kW.
This is the power output of the car engine.

M2

EXAMPLE 3

A pump raises water from a tank through a height of 3 m and expels it through a circular nozzle of radius 3 cm at $8\,\text{m}\,\text{s}^{-1}$. Find the rate at which the pump is working. Ignore any resistance forces.

In each second, the pump raises and accelerates a cylinder of water 8 m long and with radius 3 cm. The volume of this water is $V = (\pi \times 0.03^2 \times 8)\,\text{m}^3 = 0.022\,62\,\text{m}^3$

Water has a density of $1000\,\text{kg}\,\text{m}^{-3}$,
so you have mass of water $= (0.022\,62 \times 1000)\,\text{kg} = 22.62\,\text{kg}$

1 cm³ has mass 1 g

The water is raised through 3 m, so you have

$$\text{GPE gained} = (22.62g \times 3)\,\text{J} = 665\,\text{J}$$

The water is accelerated from rest to $8\,\text{m}\,\text{s}^{-1}$, so you have

$$\text{KE gained} = \left(\frac{1}{2} \times 22.62 \times 8^2\right)\text{J} = 723.8\,\text{J}$$

In each second the pump does

$$(665 + 723.8)\,\text{J} = 1388.8\,\text{J of work.}$$

So $\quad$ rate of working = 1390 W (3 s.f.)

Work done equals change of energy.

For a vehicle, the power output by the engine depends on the driving force of the engine and the velocity of the vehicle.

Suppose a car exerts a constant driving force F N so that it has a constant speed $v\,\text{m}\,\text{s}^{-1}$.

In 1 second, the car travels v m. You have

work done in 1 second = Fv J

so power output by engine = Fv W

For a vehicle travelling at $v\,\text{m}\,\text{s}^{-1}$ under a driving force F N

$$\text{power} = Fv$$

In situations where the force or the velocity is variable, this expression gives the power output at a particular instant.

EXAMPLE 4

A car is being driven at a constant speed of $20\,\text{m}\,\text{s}^{-1}$ on a level road against a constant resistance force of 260 N. Find the power output of the engine.

Let the driving force of the engine be F N.

Resolve in the direction of travel and use Newton's second law:

$F - 260 = 0$ (no acceleration)

$F = 260$ N

Therefore, you have

power = (260 × 20) W = 5200 W or 5.2 kW

EXAMPLE 5

A car of mass 800 kg is driven along a level road against a constant resistance of 450 N. The output of the engine is 7 kW. Find

a the acceleration when the speed is $10\,\text{m s}^{-1}$.

b the maximum speed of the car.

---

a Let the applied force of the engine be F N, and the acceleration be $a\,\text{m s}^{-2}$.

Power $= Fv$, so you have

$7000 = 10F$ and hence $F = 700$ N

Resolve in the direction of travel and use Newton's second law:

$700 - 450 = 800a$

and so the acceleration a is $0.3125\,\text{m s}^{-2}$

b If the car is travelling at maximum speed, acceleration is zero.

Resolve in the direction of travel:

$F - 450 = 0$

and so $F = 450$ N

Power $= Fv$, which gives

$7000 = 450v$

and so $v = 15.56$

So the maximum speed is $15.6\,\text{m s}^{-1}$ to 3 s.f.

Examples 4 and 5 show that problems involving power are usually solved using the equation

power = applied force × velocity

together with Newton's second law

resultant force = mass × acceleration

In the first equation the force referred to is the driving force (tractive force) exerted by the engine. In the second equation the force is the resultant of all the forces.

EXAMPLE 6

A car of mass 900 kg travels up a hill, inclined at 10° to the horizontal, against a constant resistance force of 250 N. Its maximum speed is $45\,\text{km h}^{-1}$. Find

a the power output of the engine

b the initial acceleration when the car reaches level road at the top of the hill, assuming that the power stays constant.

---

a The applied force of the engine is F N.

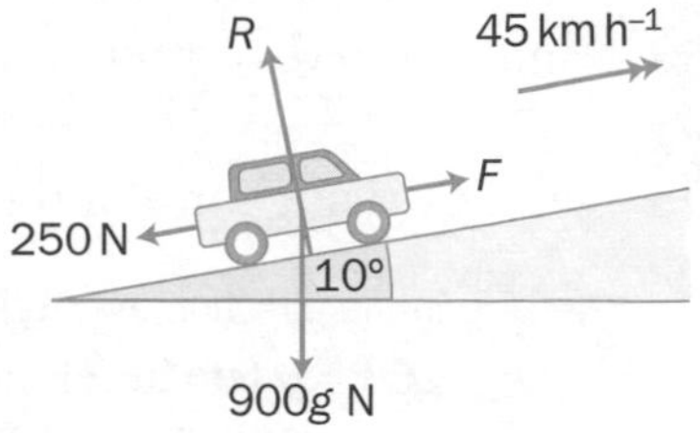

Resolve up the slope and use Newton's second law:

$F - 250 - 900g\sin 10° = 0$

so $F = 1781.6$ N

The speed is $45\,\text{km h}^{-1} = 12.5\,\text{m s}^{-1}$

So you have power $= Fv = (1781.6 \times 12.5)$ W

$= 22\,269.7$ W or 22.3 kW (to 3 s.f.)

EXAMPLE 6 (CONT.)

b When the car reaches the level road, it has the same power and, initially, the same speed. So, F is still 1781.6 N.
Let the initial acceleration be a m s^{-2}.

Resolve horizontally and use Newton's second law:

$$1781.6 - 250 = 900a$$

and so the acceleration a is 1.70 m s^{-2}

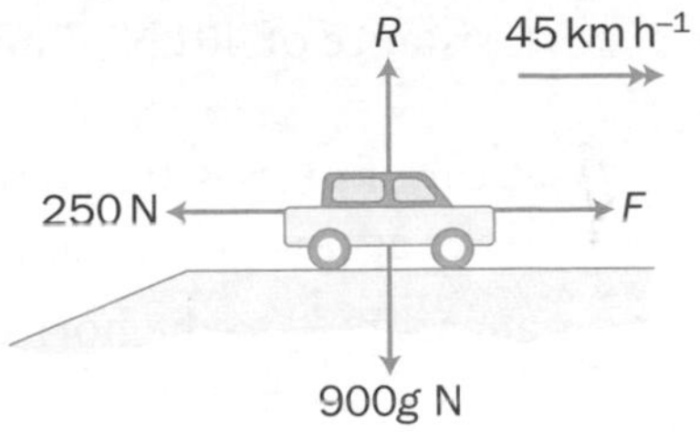

Exercise 3.3

1 A man raises a load of 20 kg through a height of 6 m using a rope and pulley. Assume the rope and pulley are light and smooth.

a What would be the man's power output if he completed the task in 30 seconds?

b If the man's maximum power output is 180 W, what is the shortest time in which he could complete the task?

2 A crate of mass 80 kg is dragged at a steady speed of 4 m s^{-1} up a slope inclined at an angle θ to the horizontal, where $\sin\theta = 0.4$. The total resistance is 250 N. Find the power required.

3 A winch has a maximum power output of 500 W. It is dragging a 200 kg crate up a slope inclined at 30° to the horizontal. The coefficient of friction between the crate and the slope is 0.6.

a Find the work done in dragging the crate a distance a m up the slope.

b Hence find the maximum speed at which the winch can drag the crate.

4 A horse, which is capable of a power output of 800 W, is able to pull a plough at a constant speed of 1.6 m s^{-1}. Find the resistance to the motion of the plough. (In fact, the traditional unit of power, 1 horsepower, is equivalent to 746 W.)

5 Arnold fills an empty washing-up liquid bottle with water and squirts a horizontal jet at a friend. The diameter of the nozzle is 4 mm and the water emerges at 10 m s^{-1}. Find his rate of working. (Assume the density of water to be 1000 kg m^{-3}.)

6 Arnold's friend gets her own back using a stirrup pump, which pumps water from a bucket. The pump raises the water through 80 cm and emits it as a jet with speed 8 m s^{-1} through a circular nozzle of radius 5 mm. Find her rate of working.

M2

7 A train has a maximum speed of $50\,\text{m s}^{-1}$ on the level against a resistance of 40 kN. Find the power output of the engine.

8 A car of mass 800 kg has a maximum speed of $75\,\text{km h}^{-1}$ up a slope against a constant resistance of 500 N. The slope is inclined at $\arcsin\frac{1}{40}$ to the horizontal. Find the power output of the engine.

9 A cyclist of mass 80 kg (including his cycle) and power 500 W is travelling on a level road against a constant resistance force of 50 N. Find his

a maximum speed

b acceleration when he is travelling at half his maximum speed.

10 A car of mass 1000 kg has a power output of 5 kW and a maximum speed of $90\,\text{km h}^{-1}$ on the level. Find

a the resistance to motion (assumed to be constant)

b the car's maximum speed up a slope inclined at $\arcsin\frac{1}{25}$ to the horizontal.

11 A train of mass 40 tonnes has a maximum speed of $15\,\text{m s}^{-1}$ up a slope against a resistance of 50 kN. The slope is inclined to the horizontal at an angle θ, where $\sin\theta = \frac{1}{50}$. Assuming the resistance is constant, find the maximum speed of the train down the same slope.

12 The frictional resistances acting on a train are $\frac{1}{100}$ of its weight. Its maximum speed up an incline of $\arcsin\frac{1}{80}$ is $48\,\text{km h}^{-1}$. Find its maximum speed on the level.

13 An open truck of mass 5 tonnes is carrying a load of 500 kg of fish up a hill against a constant resistance of 800 N. It is travelling at its maximum speed of $10\,\text{m s}^{-1}$. The hill is inclined at $\arcsin\frac{1}{50}$ to the horizontal. A flock of gulls, mass 1000 kg, descends on the lorry to eat the fish.

a Find the initial deceleration of the truck and the new maximum speed.

As the truck reaches the top of the hill and moves on to a level road, the gulls, having eaten all the fish, fly away.

b Find the initial acceleration of the truck.

14 A train of mass 50 tonnes has a maximum speed on the level of $50\,\text{km}\,\text{h}^{-1}$ when the engine is working at 80 kW. Assuming that resistance is constant, how far would the train travel before coming to rest if, when travelling at maximum speed, the engine were disengaged and the train allowed to coast?

15 **a** A car of mass 900 kg pulls a trailer of mass 200 kg. The resistance to motion of the car is 200 N and of the trailer is 80 N. Find the power output of the engine if the maximum speed on the level is $40\,\text{m}\,\text{s}^{-1}$.

b The car and trailer are travelling at $8\,\text{m}\,\text{s}^{-1}$ on a hill, inclined at $\arcsin \frac{1}{40}$ to the horizontal. If the resistance is constant and the engine is exerting full power, find

i the acceleration

ii the tension in the towbar.

16 The resistance to motion of a car is proportional to its speed. A car of mass 1000 kg has a maximum speed of $45\,\text{m}\,\text{s}^{-1}$ on the level when its power output is 8 kW. Find its acceleration when it is travelling on the level at $20\,\text{m}\,\text{s}^{-1}$ and its engine is working at 6 kW.

17 **a** A lorry of mass 10 tonnes has a maximum speed of $20\,\text{m}\,\text{s}^{-1}$ up a slope inclined at $\arcsin \frac{1}{100}$ to the horizontal when its engine is working at 70 kW. Find the resistance to motion.

b If the resistance is proportional to the square of the speed, find the maximum speed of the lorry on the level when the engine is working at the same rate.

18 A car has a maximum power P. The resistance to motion is kv. Its maximum speed up a certain slope is V and its maximum speed down the same slope is $2V$. Show that $V = \sqrt{\frac{P}{2k}}$

19 A cyclist and her cycle have a combined mass of 80 kg. The resistance to motion is proportional to the speed. On the level, she can travel at a maximum speed of $10\,\text{m}\,\text{s}^{-1}$, and she can freewheel down an incline of angle θ at $14\,\text{m}\,\text{s}^{-1}$. Find the maximum speed at which she can go up the same incline.

1 A packing case of mass 30 kg is dragged at constant speed across a rough horizontal floor by a horizontal rope. The coefficient of friction is 0.5.

a Find the work done by the tension in the rope in moving the packing case a distance of 8 m.

b Repeat part **a** with the rope inclined at 20° to the horizontal.

2 A constant horizontal force of 30 N is applied to a particle of mass 0.6 kg which is at rest at a point *A* on a smooth horizontal surface. The particle passes a point *B* with speed $12\,\text{m}\,\text{s}^{-1}$.

a Find the kinetic energy gained by the particle in moving from *A* to *B*.

b State the work done by the force.

c Find the distance *AB*.

3 Mary is standing at the top of a tower. She leans over the edge and throws a ball of mass 0.1 kg vertically upwards at a speed of $4\,\text{m}\,\text{s}^{-1}$ from a point 12 m above the ground, so that it travels upwards and then falls to the ground.

a Calculate the kinetic energy of the ball as it leaves Mary's hand.

b Use an energy method to find how far the ball rises from its starting point.

c Calculate the kinetic energy of the ball when it hits the ground.

d Find the speed with which the ball hits the ground.

4 A particle of mass 2 kg is dropped from rest at a point 6 m above the ground.

a Assuming there is no air resistance, use an energy method to find the speed with which it hits the ground.

b In fact there is a constant resistance force, *R* N, and the particle hits the ground at $10\,\text{m}\,\text{s}^{-1}$. Find the value of *R*.

5 A particle of mass 10 kg is sliding down a rough plane inclined at an angle θ to the horizontal, where $\sin\theta = \frac{3}{5}$. It passes a point *A* travelling at $2\,\text{m}\,\text{s}^{-1}$ and a point *B* at $8\,\text{m}\,\text{s}^{-1}$. $AB = 10\,\text{m}$.

a Find the energy lost by the particle in travelling from *A* to *B*.

b State the work done against friction.

c Hence find the coefficient of friction between the plane and the particle.

6 A car of mass 1200 kg has a maximum speed of 180 km h^{-1} on a level road. The power output of the engine is 60 kW.

a Calculate the resistance force on the car.

The car then travels up a slope inclined at 10° to the horizontal. At the start of the slope it is travelling at 54 km h^{-1}. The power and the resistance are unchanged.

b Calculate the acceleration of the car at the start of the slope.

c Calculate the maximum speed of the car on the slope.

7 A car of mass 1000 kg has a maximum speed of 40 m s^{-1} on a horizontal road. The total resistance force is 2800 N.

a Find the power output of the engine.

The power and resistance are constant.

b Calculate the acceleration of the car at a moment when it is travelling on the level at 20 m s^{-1}.

c Find the maximum speed of the car when travelling down a slope inclined at 5° to the horizontal.

8 A cyclist has a maximum speed of 8 m s^{-1} on a horizontal road against a constant resistance force of 25 N. The rider and her machine have a total mass of 90 kg.

a Calculate the power output of the cyclist.

When travelling up a hill she must maintain a speed of at least 0.5 m s^{-1} to avoid wobbling dangerously.

b Assuming the power and the resistance remain the same, find the angle of the steepest slope up which she can safely cycle.

9 A car of mass 1000 kg is travelling on a level road at a constant speed of 108 km h^{-1}. The power output of the engine is 60 kW.

a Calculate the magnitude of the resistance force acting on the car.

The car reaches the start of a downhill stretch of road inclined at $\arcsin \frac{1}{10}$ to the horizontal. The power and resistance are unchanged.

b Calculate the initial acceleration of the car.

c Calculate the maximum speed of the car down the slope.

3

Exit

Summary

Refer to

- If the point of application of a force F undergoes a displacement s in the direction of F, then
  work done by $F = F \times s$ joules
- If the point of application of a force F undergoes a displacement s at an angle θ to the direction of F, then
  work done by $F = Fs\cos\theta$ 3.1
- For an object of mass m kg at a height h m from the zero level
  gravitational potential energy (GPE) $= mgh$ J
- An object of mass m kg travelling at $v\,\text{m s}^{-1}$ has
  kinetic energy (KE) $= \frac{1}{2}mv^2$ J
- The principle of conservation of mechanical energy states that the total mechanical energy of a system remains constant provided no external work is done and there are no sudden changes in the motion of the system.
- The work–energy principle states that the total work done on a system equals the change in mechanical energy of the system. 3.2
- Power is the rate at which work is done. The SI unit of power is the watt (W), where 1 W = 1 joule per second.
- For a vehicle travelling at $v\,\text{m s}^{-1}$ under a driving force F N
  Power $= Fv$ 3.3

Links

Wind turbines work by converting kinetic energy in the wind into mechanical energy in the blades which is then converted into electrical energy.

Engineers use aerodynamical modelling to determine the optimum tower height, control systems, number of blades as well as the shape, size and weight of the blades to convert the energy most efficiently.

Knowledge of the usual wind speeds and hence the available kinetic energy in a particular area enables the engineers to assess the suitability of possible wind farm locations.

4 Collisions

This chapter will show you how to
- use Newton's experimental law of restitution to calculate the outcome of collisions
- analyse collisions between particles and between a particle and a wall
- find the loss of kinetic energy during a collision.

Before you start

You should know how to:	Check in:
1 Solve linear simultaneous equations.	1 Solve the equations $3u + 2v = 10$ $u - v = 5$
2 Calculate kinetic energy.	2 Calculate the kinetic energy of a particle of mass 3 kg travelling with speed $4\,\text{m s}^{-1}$.
3 Recall the equations of motion with constant acceleration.	3 State the equations of motion with constant acceleration.

4.1 Impulse and momentum – a reminder

- If a force $\mathbf{F}$ N acts for a time t s, it exerts an **impulse** of $\mathbf{I} = \mathbf{F}t$ N s.
- Impulse is a vector quantity.
- If a particle of mass m kg has velocity $\mathbf{v}$ m s^{-1}, it has **momentum** $m\mathbf{v}$ N s.
- Momentum is a vector quantity.
- Impulse = change of momentum; that is, for a particle of mass m with initial velocity $\mathbf{u}$ and final velocity $\mathbf{v}$, you have $\mathbf{I} = m\mathbf{v} - m\mathbf{u}$
- The **principle of conservation of momentum**: If no external force acts on a system, the total momentum of the system remains constant.

See M1 for more revision of impulse and momentum.

EXAMPLE 1

A particle of mass 3 kg is moving with velocity $(2\mathbf{i} + 3\mathbf{j})$ m s^{-1} when it receives an impulse of $(3\mathbf{i} - 6\mathbf{j})$ N s.
Find its new velocity.

Let the new velocity be $\mathbf{v}$ m s^{-1}.
Then $3\mathbf{i} - 6\mathbf{j} = 3\mathbf{v} - 3(2\mathbf{i} + 3\mathbf{j})$

$$3\mathbf{v} = 9\mathbf{i} + 3\mathbf{j}$$

so the new velocity v is $(3\mathbf{i} + \mathbf{j})$ m s^{-1}

Using impulse = change of momentum

EXAMPLE 2

Particles A and B are travelling along the same straight line. A has mass 2 kg and speed 5 m s^{-1}. B has mass 3 kg and speed 2 m s^{-1}. They collide and coalesce (stick together).

a Find their velocity after the collision if they were originally travelling
  i in the same direction **ii** in opposite directions.

b Find the loss of kinetic energy in each case.

---

a i Momentum before the collision:
Particle A: (2×5) N s = 10 N s,
Particle B: (3×2) N s = 6 N s. Let v be the velocity of the combined particles A and B after the collision.
Combined momentum after collision = $5v$ N s

Use conservation of momentum:

$$5v = 10 + 6 = 16 \quad \text{so} \quad v = 3.2$$

So the velocity is 3.2 m s^{-1} in the same direction.

ii The only change is that B's velocity is -2 m s^{-1} and hence momentum $(3 \times (-2))$ N s = -6 N s

$$5v = 10 - 6 = 4 \quad \text{so} \quad v = 0.8$$

So the velocity is 0.8 m s^{-1} in A's original direction.

b Initial KE = $\left(\frac{1}{2} \times 2 \times 5^2 + \frac{1}{2} \times 3 \times 2^2\right)$ J = 31 J

In **i**, final KE = $\left(\frac{1}{2} \times 5 \times 3.2^2\right)$ J = 25.6 J, a loss of 5.4 J

In **ii**, final KE = $\left(\frac{1}{2} \times 5 \times 0.8^2\right)$ J = 1.6 J, a loss of 29.4 J

Sketch before and after diagrams for each situation.

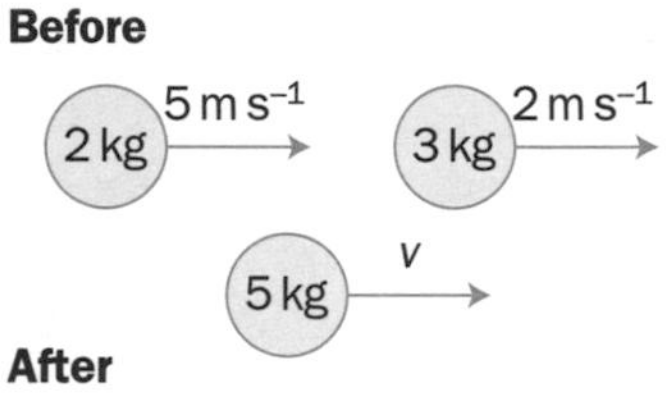

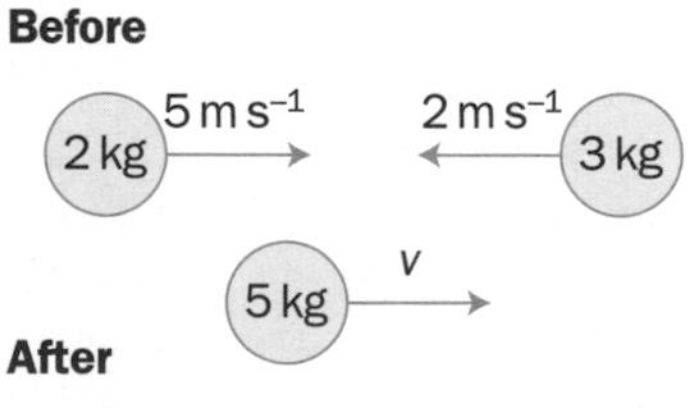

Energy is not conserved if there are sudden changes, such as collisions.

Exercise 4.1

1 A body of mass 5 kg is travelling with velocity $(2\mathbf{i} + 3\mathbf{j})\ \text{m s}^{-1}$ when it receives an impulse $\mathbf{I}$ N s. This changes its velocity to $(4\mathbf{i} + 7\mathbf{j})\ \text{m s}^{-1}$. Find $\mathbf{I}$ and show that it has a magnitude of $10\sqrt{5}$ N s.

2 Particles A and B are travelling along the same straight line. A has mass 1 kg and speed $4\ \text{m s}^{-1}$. B has mass 3 kg and speed $3\ \text{m s}^{-1}$. They collide and coalesce.

a Find their combined velocity if they were originally travelling

i in the same direction

ii in opposite directions.

b Find the loss of kinetic energy in each case.

3 A particle A, of mass 3 kg, is travelling in a straight line at $5\ \text{m s}^{-1}$ when it collides head-on with particle B, of mass 2 kg, travelling in the opposite direction at $4\ \text{m s}^{-1}$. After the collision A is moving in the same direction but with its speed reduced to $1\ \text{m s}^{-1}$.

a Find the velocity of B after the collision.

b Calculate the impulse exerted by A on B during the collision.

c Calculate the kinetic energy lost by the system during the collision.

d Explain why particle A could not have been moving in the same direction at $2\ \text{m s}^{-1}$ after the collision.

4 A particle A, of mass 4 kg, travelling at $6\ \text{m s}^{-1}$ catches up and collides with a particle B, of mass m kg, travelling at $1\ \text{m s}^{-1}$. The particles coalesce and move with speed $3\ \text{m s}^{-1}$.

a Calculate the value of m.

b What would their common velocity have been if they had been travelling in opposite directions before the collision?

5 A bullet of mass m is fired at a block of wood of mass M which is suspended on the end of a light string. The bullet enters the block horizontally with speed v and becomes embedded. The block swings until it reaches a height h above its original level.

Show that $h = \dfrac{m^2v^2}{2g(m + M)^2}$

M2

4.2 Newton's experimental law of restitution

When two objects collide and rebound, there are two unknown quantities – the two post-collision velocities – and so you need two equations to solve the problem.

One equation comes from conservation of momentum.
The second equation depends on the 'elasticity' of the objects.

The **approach speed** is the rate at which the gap between the objects decreases before the collision.
The **separation speed** is the rate at which the gap between the objects increases after the collision.

Newton observed that separation speed $\leqslant$ approach speed, and that for a given pair of objects the ratio between these speeds is a constant, e, the **coefficient of restitution**.

> Newton's experimental law of restitution
>
> $$\frac{\text{separation speed}}{\text{approach speed}} = e \text{ (a constant)}$$

If $e = 0$ the impact is **inelastic**. No rebounding occurs.
If $e = 1$ the impact is **perfectly elastic**.

In practice, perfect elasticity does not occur, so in the real world $0 \leqslant e < 1$

EXAMPLE 1

A particle of mass 3 kg moving at $6\,\text{m s}^{-1}$ collides directly with a stationary particle of mass 5 kg. The coefficient of restitution between the particles is 0.4. Find their velocities after impact.

Let the velocities after impact be u and v.

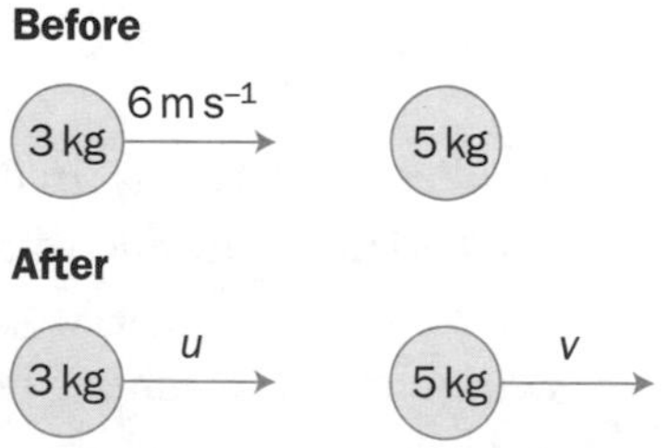

Both particles are free to move, so momentum is conserved.

momentum before impact $= 3 \times 6 = 18$

momentum after impact $= 3u + 5v$

So you have $\quad 3u + 5v = 18 \quad$ [1]

approach speed $= 6\,\text{m s}^{-1}$

separation speed $= v - u$

Use Newton's law of restitution: $\quad \dfrac{v - u}{6} = 0.4$

So you have $\quad v - u = 2.4 \quad$ [2]

Solve [1] and [2]: $\quad u = 0.75\,\text{m s}^{-1}$, $v = 3.15\,\text{m s}^{-1}$

Always sketch before and after diagrams.

EXAMPLE 2

A particle *A*, of mass 3 kg travelling at 6 m s^{-1}, collides with a particle *B*, of mass 2 kg travelling in the same direction at 2 m s^{-1}.
The coefficient of restitution between the particles is 0.8.
Find their velocities immediately after impact.

Let the velocities after impact be *u* and *v*, as shown.

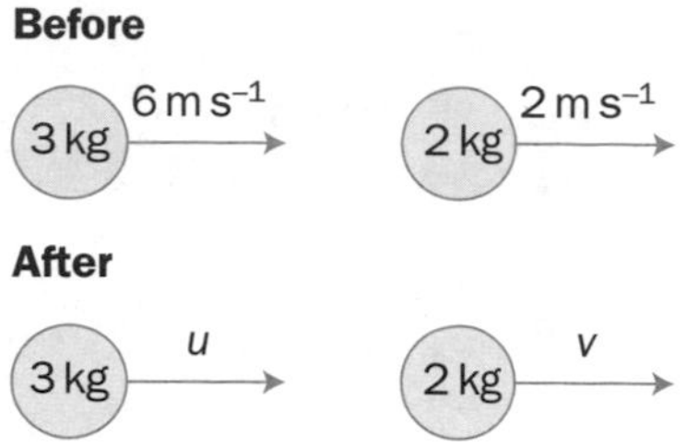

Use conservation of momentum:

$$3u + 2v = 3 \times 6 + 2 \times 2$$

and so $\quad 3u + 2v = 22 \qquad [1]$

approach speed $= 6 - 2 = 4$
separation speed $= v - u$

Use Newton's law of restitution: $\dfrac{v-u}{4} = 0.8$

So you have $\quad v - u = 3.2 \qquad [2]$

Solve [1] and [2]: $\quad u = 3.12\,\text{m s}^{-1}, v = 6.32\,\text{m s}^{-1}$

EXAMPLE 3

Two identical particles collide head-on with speeds 5 m s^{-1} and 3 m s^{-1}. The coefficient of restitution between the particles is 0.5.
Find the velocities of the particles immediately after impact.

Let the particles have mass *m* and the velocities after impact be *u* and *v*, as shown.

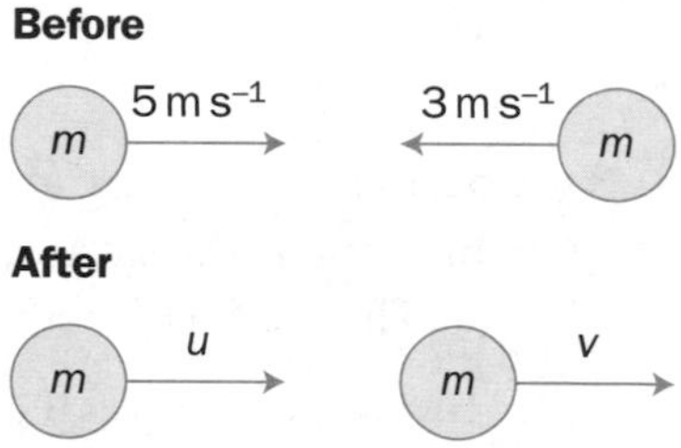

Use conservation of momentum:

$$mu + mv = 5m - 3m$$

and so $\quad u + v = 2 \qquad [1]$

approach speed $= 5 - (-3) = 8$

separation speed $= v - u$

Use Newton's law of restitution: $\dfrac{v-u}{8} = 0.5$

So you have $\quad v - u = 4 \qquad [2]$

Solve [1] and [2]: $\quad u = -1\,\text{m s}^{-1}, v = 3\,\text{m s}^{-1}$

So the particles move in opposite directions with speeds 1 m s^{-1} and 3 m s^{-1}.

Always define the unknown velocities *u* and *v* in the same direction, as in the diagram. The separation speed is then always $(v - u)$. If either particle is in fact moving in the opposite direction, the velocity you calculate will be negative, as in Example 3.

M2

Kinetic energy is always lost during a collision. The smaller the value of *e*, the greater the energy loss will be.

A perfectly elastic collision ($e = 1$) would involve no energy loss.
In practice some energy is always lost in the form of heat and sound.

Exercise 4.2

1 Particles A and B collide directly. In each case the diagram shows the velocities before and after impact. Calculate the value of e, the coefficient of restitution.

Before **After**

a
A $8\,\mathrm{m\,s^{-1}}$ B At rest A At rest B $6\,\mathrm{m\,s^{-1}}$

b
A $10\,\mathrm{m\,s^{-1}}$ B $4\,\mathrm{m\,s^{-1}}$ A $3\,\mathrm{m\,s^{-1}}$ B $6\,\mathrm{m\,s^{-1}}$

c
A $5\,\mathrm{m\,s^{-1}}$ $4\,\mathrm{m\,s^{-1}}$ B $2\,\mathrm{m\,s^{-1}}$ A B $3\,\mathrm{m\,s^{-1}}$

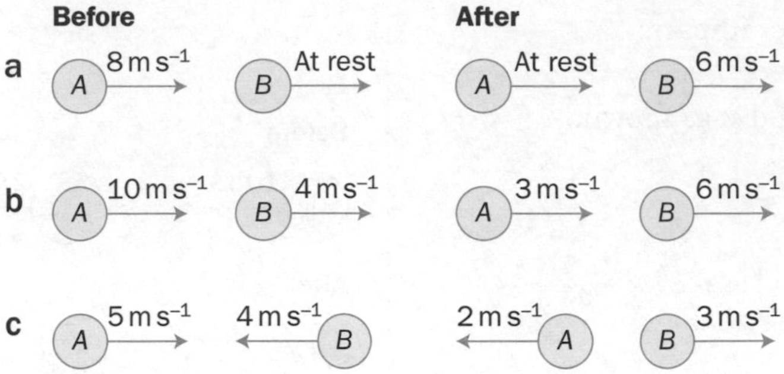

2 The table shows the mass and initial velocity of two particles, A and B, and the coefficient of restitution. In each case calculate the velocities of the particles after the collision.

	A		B		
	Mass (kg)	Initial velocity (m s^{-1})	Mass (kg)	Initial velocity (m s^{-1})	e
a	4	8	3	0	0.6
b	2	6	3	2	0.4
c	5	4	2	−2	0.8
d	2	5	5	−4	0.2

3 A particle A, moving at $3\,\mathrm{m\,s^{-1}}$, collides with a second particle, B, moving in the same direction at $1\,\mathrm{m\,s^{-1}}$. B has twice the mass of A. The coefficient of restitution is 0.5. Find

   a the velocity of each particle immediately after the collision

   b the fraction of kinetic energy lost in the collision.

4 A particle A, of mass $3\,\mathrm{kg}$ and travelling at $10\,\mathrm{m\,s^{-1}}$, collides with a second particle, B, of mass $5\,\mathrm{kg}$, travelling in the same direction at $2\,\mathrm{m\,s^{-1}}$. After the collision, A is moving in the same direction with its speed reduced to $4\,\mathrm{m\,s^{-1}}$. Calculate

   a the new speed of B

   b the coefficient of restitution.

5 Particles A and B, of mass $5\,\mathrm{kg}$ and $3\,\mathrm{kg}$ respectively, are travelling towards each other along the same straight line. Their speeds are $3\,\mathrm{m\,s^{-1}}$ and $2\,\mathrm{m\,s^{-1}}$ respectively. Particle A is brought to rest by the collision. Calculate

   a the coefficient of restitution between the particles

   b the kinetic energy lost during the collision.

6 A particle of mass 6 kg, travelling at 8 m s^{-1}, is brought to rest in a collision with a second particle of mass 4 kg. If $e = 0.3$, calculate the initial and final velocities of the second particle.

7 Particles A and B, of mass m and $2m$ respectively, are travelling towards each other along the same straight line, each with a speed of u. The coefficient of restitution is 0.5. Calculate the kinetic energy lost when the particles collide.

8 Two perfectly elastic particles of equal mass are travelling in opposite directions along the same straight line.

a Show that when they collide they exchange velocities.

b Would the same be true if they were initially travelling in the same direction?

9 Particles A and B have masses of 3 kg and 2 kg respectively. They are connected together by a light inextensible string. The particles lie at rest on a smooth horizontal surface. The coefficient of restitution between the particles is 0.5. A is projected towards B with velocity 10 m s^{-1}.

See M1 for revision.

a Calculate

i the velocities of the particles after the collision

ii the common velocity of the particles after the string becomes taut

iii the kinetic energy lost during the whole process.

b Explain why the answers to **a** parts **ii** and **iii** are independent of e.

10 Particle A of mass m, moving at 7 m s^{-1} collides with particle B, of mass km, moving in the same direction at 1 m s^{-1}. After the collision the speed of B is twice the speed of A. The coefficient of restitution is 0.75. Find the two possible values of k.

11 Particles A and B have equal mass. A is at rest at the bottom of a smooth curved ramp. B is released from rest at a point on the ramp 2.5 m higher than A, as shown. The coefficient of restitution between the particles is $\frac{3}{7}$. After the collision the particles move on a rough horizontal surface, with coefficient of friction 0.4. Find the distance between the particles when they have both come to rest.

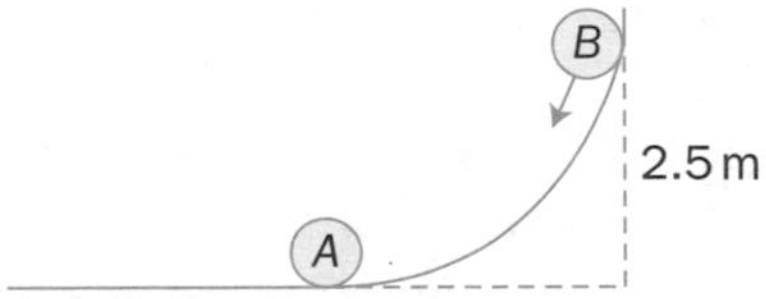

12 Marbles A and B, of equal mass, rest at the opposite ends of a diameter of a smooth horizontal circular groove. A is projected along the groove and strikes B after 2 s. $e = 0.5$.

a How long will it be before B strikes A?

b How many collisions will there be in the first 5 minutes?

4.3 Collision with a wall

Newton's experimental law of restitution still applies in the case of a particle colliding with a fixed obstacle, such as a wall. That is

$$\frac{\text{speed of particle leaving wall}}{\text{speed of particle approaching wall}} = e$$

However, because the wall is not free to move there are external forces acting on the system. Hence momentum is not conserved.

You will only need to deal with situations where the particle is travelling at right angles to the wall.

EXAMPLE 1

A particle of mass 4 kg travelling at $8\,\text{m}\,\text{s}^{-1}$ strikes a fixed wall at right angles and rebounds. The coefficient of restitution is 0.3. Find

a the velocity of the particle after impact

b the impulse exerted on the wall.

---

a Let the separation speed be v. You have

$$\frac{v}{8} = 0.3 \quad \text{so} \quad v = 2.4\,\text{m}\,\text{s}^{-1}$$

So, taking left to right in the diagram as the positive direction, the velocity after impact is $-2.4\,\text{m}\,\text{s}^{-1}$.

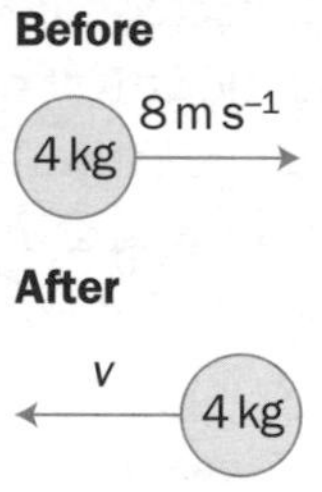

b Momentum of particle before impact $= (4 \times 8)\,\text{N s} = 32\,\text{N s}$

Momentum of particle after impact $= (4 \times (-2.4))\,\text{N s}$

$= -9.6\,\text{N s}$

Change of momentum $= ((-9.6) - 32)\,\text{N s} = -41.6\,\text{N s}$

The particle receives an impulse of $-41.6\,\text{N s}$, so by Newton's third law the wall receives an impulse of $41.6\,\text{N s}$.

EXAMPLE 2

A ball of mass m is dropped from rest onto horizontal ground, striking it at $14\,\text{m}\,\text{s}^{-1}$. The coefficient of restitution is 0.5. Ignoring air resistance, find

a the time which elapses between the first and the second bounce

b the total time between the first bounce and when the ball comes finally to rest.

---

a For the first bounce, approach speed $= 14\,\text{m}\,\text{s}^{-1}$.

Separation speed is V_1, where $\frac{V_1}{14} = 0.5 \quad \text{so} \quad V_1 = 7\,\text{m}\,\text{s}^{-1}$

The ball leaves the ground at $7\,\text{m}\,\text{s}^{-1}$.

Taking upwards as positive, you have

$u = 7$, $a = -9.8$, $s = 0$ (when the ball returns to the ground)

EXAMPLE 2 (CONT.)

Use $s = ut + \frac{1}{2}at^2$: $0 = 7t - 4.9t^2$ so $t = \frac{10}{7}$ (or 0)

So there is an interval of $\frac{10}{7}$ s between the first and second bounces.

b For the second bounce, approach speed $= 7\,\text{m s}^{-1}$

Separation speed is V_2, where $\frac{V_2}{7} = 0.5$ so $V_2 = 3.5\,\text{m s}^{-1}$

Use $s = ut + \frac{1}{2}at^2$ as before:

$0 = 3.5t - 4.9t^2$ so $t = \frac{5}{7}$ (or 0)

So there is an interval of $\frac{5}{7}$ s between the second and third bounces.

Continuing, you can see that after each bounce the ball stays in the air for half the time of the previous stage.

See C2 for revision of series.

Total time $= \left(\frac{10}{7} + \frac{5}{7} + \frac{5}{14} + \frac{5}{28} + \cdots\right)$ s

This is an infinite geometric series with $r = \frac{1}{2}$, for which

$$S_\infty = \frac{\frac{10}{7}}{1 - \frac{1}{2}} = \frac{20}{7}$$

The sum to infinity of a geometric series with first term a and common ratio r is $S_\infty = \frac{a}{1-r}$

So the total time $= \frac{20}{7}$ s

EXAMPLE 3

Particles A and B, of mass 1 kg and 3 kg respectively, lie at rest on a smooth horizontal plane. The line AB is perpendicular to a vertical wall. The coefficient of restitution between the particles is 0.4. A is projected towards B at $10\,\text{m s}^{-1}$. After the collision B hits the wall and rebounds. The coefficient of restitution between B and the wall is e.

Show that the particles will collide again if $e > \frac{1}{7}$.

When the particles collide, momentum is conserved, so

$u + 3v = 10$ [1]

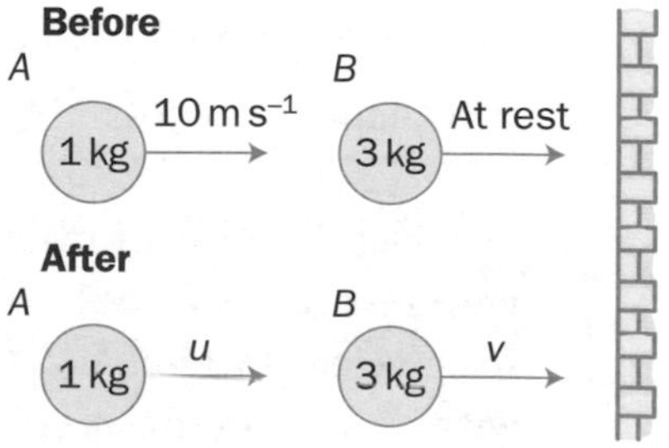

Apply Newton's experimental law:

$\frac{v - u}{10} = 0.4$ so $v - u = 4$ [2]

Solve [1] and [2]: $u = -0.5$ and $v = 3.5$

So, A is moving away from the wall at $0.5\,\text{m s}^{-1}$ and B is moving towards the wall at $3.5\,\text{m s}^{-1}$.

Example 3 is continued on the next page.

M2

EXAMPLE 3 (CONT.)

B hits the wall and rebounds with speed $V\,\mathrm{m\,s^{-1}}$.

Use Newton's experimental law: $\frac{V}{3.5} = e$ so $V = 3.5\,e$

B will overtake and collide with A provided

$3.5e > 0.5$ that is if $e > \frac{1}{7}$

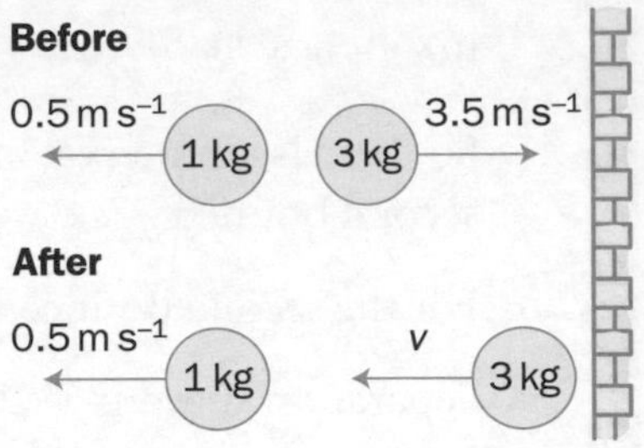

Exercise 4.3

1 A particle collides normally (at right angles) with a fixed wall. It approaches the wall with speed $u\,\mathrm{m\,s^{-1}}$ and rebounds with speed $v\,\mathrm{m\,s^{-1}}$. The coefficient of restitution is e. Find

a e if $u = 6$ and $v = 2$

b v if $u = 8$ and $e = 0.7$

c u if $v = 6$ and $e = 0.4$

2 A particle of mass 2 kg is at rest at a point O on a smooth horizontal surface halfway between two vertical walls which are 8 m apart. The particle is projected at $4\,\mathrm{m\,s^{-1}}$ directly at one of the walls. The coefficient of restitution between the particle and each wall is 0.5.

a Find how long it takes the particle to strike each wall and return to O.

b Calculate the amount of kinetic energy lost by the particle during this process.

3 A ball is dropped from a height of 3.6 m on to a horizontal surface. It rebounds to a height of 0.4 m. Calculate

a the speed at which the ball hits the surface

b the speed at which the ball leaves the surface

c the coefficient of restitution between the ball and the surface.

The ball continues to bounce repeatedly.

d Calculate the total distance travelled by the ball before coming to rest.

4 Particles A and B, of mass 4 kg and 2 kg respectively, are moving with respective speeds of $1\,\mathrm{m\,s^{-1}}$ and $10\,\mathrm{m\,s^{-1}}$ directly towards a fixed vertical wall. B hits the wall, rebounds and then collides with A. The coefficient of restitution between the wall and particle B is 0.4 and between the particles is 0.2.

a Show that B is brought to rest when it collides with A.

b Calculate the speed at which A is travelling after the collision.

M2

5 A particle A, mass 2 kg and velocity $10\,\mathrm{m\,s^{-1}}$, is moving on a smooth horizontal surface. It overtakes and collides with a second particle B, mass 1 kg and velocity $5\,\mathrm{m\,s^{-1}}$. B then impacts normally with a vertical wall. The coefficient of restitution between A and B is 0.5, and between B and the wall is 0.75.

a Find the velocities of A and B after they collide.

b Find the velocities of the particles after they collide for a second time.

c Find what happens after this second impact.

6 The diagram shows two particles, A of mass 1 kg and B of mass 5 kg, moving at $6\,\mathrm{m\,s^{-1}}$ and $4\,\mathrm{m\,s^{-1}}$ respectively on a smooth horizontal surface directly towards a vertical wall. The coefficient of restitution between A and B is 0.2, and between B and the wall is 0.5.

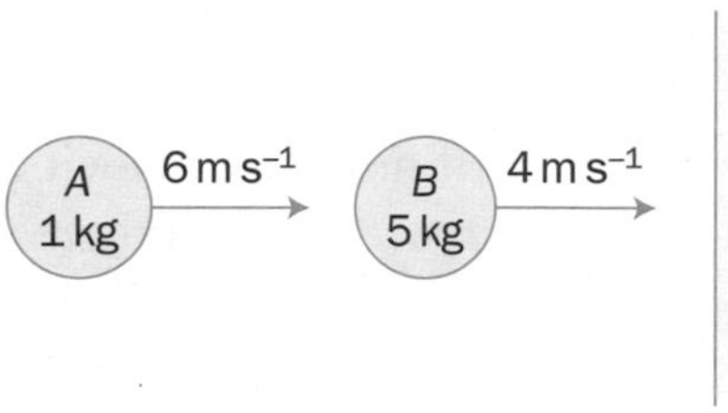

a Find the velocities of the two particles after all impacts have taken place if the first impact is between
 i A and B ii B and the wall.

b Find in each case the total loss of kinetic energy.

7 The ceiling in a sports hall is horizontal and 10 m above the floor. A ball is projected vertically upwards from a point 2 m above the floor with speed u. It strikes the ceiling, then the floor and just reaches the ceiling for a second time. If the coefficient of restitution between both the ceiling and floor and the ball is 0.5, find u.

8 Two parallel vertical walls stand a distance $2d$ apart on a smooth horizontal surface. Identical particles A and B are at rest on the surface halfway between the walls. The coefficient of restitution between the particles and the first wall is e_1, and between the particles and the second wall is e_2, where $e_1 > e_2$. The particles are set in motion simultaneously with speed u, particle A towards wall 1 and particle B towards wall 2. Each particle strikes both walls and arrives back at the middle.

a Show that they each have the same speed at the end of the process.

b Show that A finishes first, and find an expression for the difference in their travel times.

c If the surface is rough, with coefficient of friction μ, show that both particles complete the journey provided

$$u \geqslant \frac{\sqrt{2\mu g d\left(e_1^2 e_2^2 + 2e_1^2 + 1\right)}}{e_1 e_2}$$

4.4 Collisions with three particles

Problems involving three particles moving in the same straight line do not involve any new mathematics, but you need to use clear notation for the unknown velocities to avoid confusion. One option is to always use the same letter for a given particle with a subscript to show which impact is being considered.

You will not be expected to deal with more than three particles.

EXAMPLE 1

Particles A, B and C lie at rest in a straight line in that order. Their masses are 3 kg, 2 kg and 4 kg respectively. A is projected towards B with velocity $8\,\text{m s}^{-1}$. The coefficient of restitution in each impact is 0.6.

a Show that there will be at least three collisions.

b By finding the velocities of the particles after the third impact decide if there will be further collisions.

Use u, v and w for velocities of A, B and C respectively, with subscripts to show first, second and third impacts.

a First impact (A and B)

Use conservation of momentum:

$$3u_1 + 2v_1 = 24 \qquad [1]$$

Use Newton's experimental law:

$$v_1 - u_1 = 4.8 \qquad [2]$$

Solve [1] and [2]:

$$u_1 = 2.88\,\text{m s}^{-1},\ v_1 = 7.68\,\text{m s}^{-1}$$

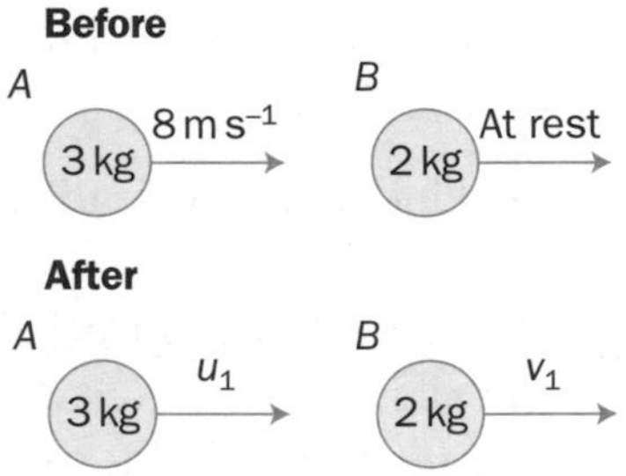

Second impact (B and C)

Use conservation of momentum:

$$2v_2 + 4w_2 = 15.4 \qquad [3]$$

Use Newton's experimental law:

$$w_2 - v_2 = 4.61 \qquad [4]$$

Solve [3] and [4]:

$$v_2 = -0.512\,\text{m s}^{-1},\ w_2 = 4.10\,\text{m s}^{-1}$$

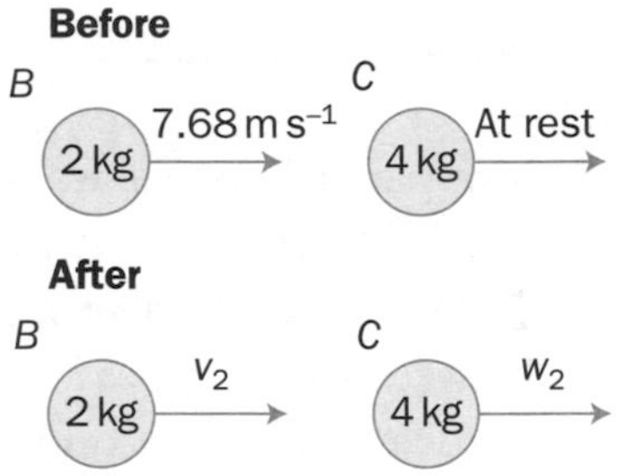

At this stage A has velocity $2.88\,\text{m s}^{-1}$ and B has velocity $-0.512\,\text{m s}^{-1}$, so there is at least one more collision.

b Third impact (A and B)

Use conservation of momentum:

$$3u_3 + 2v_3 = 7.62 \qquad [5]$$

Use Newton's experimental law:

$$v_3 - u_3 = 2.04 \qquad [6]$$

Solve [5] and [6]:

$$u_3 = 0.709\,\text{m s}^{-1},\ v_3 = 2.74\,\text{m s}^{-1}$$

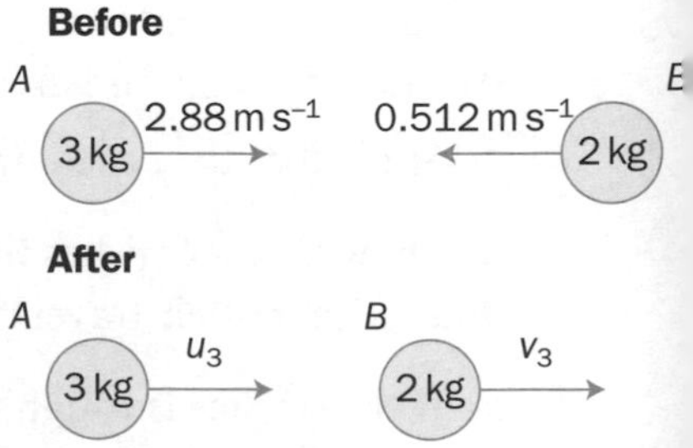

At this stage B has velocity $2.74\,\text{m s}^{-1}$ and C has velocity $4.10\,\text{m s}^{-1}$, so B will not catch up with C.

There will be no more collisions.

All values have been shown correct to 3 s.f., but you should work with more accuracy than this. Store each value in a separate memory on your calculator, if possible.

Exercise 4.4

1 Particles A, B and C, each of mass m, lie at rest in a straight line on a smooth horizontal surface in the order stated. A is projected directly towards B with velocity $10\,\text{m s}^{-1}$. The coefficient of restitution in each impact is 0.5. Show that there will be three collisions and find the final velocities of the particles.

2 Particles A, B and C have masses of 3 kg, 2 kg and 1 kg respectively. They are moving, in that order, along a straight line with velocities of $3\,\text{m s}^{-1}$, $2\,\text{m s}^{-1}$ and $1\,\text{m s}^{-1}$ respectively. The collision between A and B, which happens first, is perfectly elastic. The second collision, between B and C, has coefficient of restitution e. Show that there will be no more collisions if $e < \frac{4}{11}$.

3 The diagram shows three particles, A, B and C, moving in the same straight line. They have masses of 3 kg, 4 kg and 2 kg respectively, and move at speeds of $4\,\text{m s}^{-1}$, $2\,\text{m s}^{-1}$ and $1\,\text{m s}^{-1}$ with B moving in the opposite direction from A and C. The coefficient of restitution is 0.5 in all impacts.

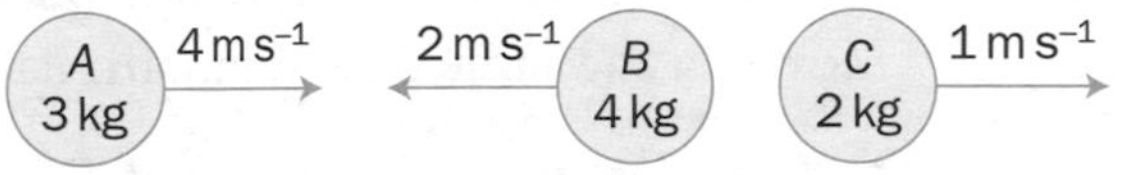

a Show that there are just two collisions.

b Find the total loss of kinetic energy.

4 The diagram shows particles A, B and C of masses 1 kg, 2 kg and 2 kg respectively. Initially they are at rest in that order in a straight line. A and B are connected by a light inextensible string which is slack. B is then propelled towards C at $6\,\text{m s}^{-1}$. The coefficient of restitution between B and C is 0.5, as is the coefficient of restitution between A and B. The string becomes taut before B reaches C. Find

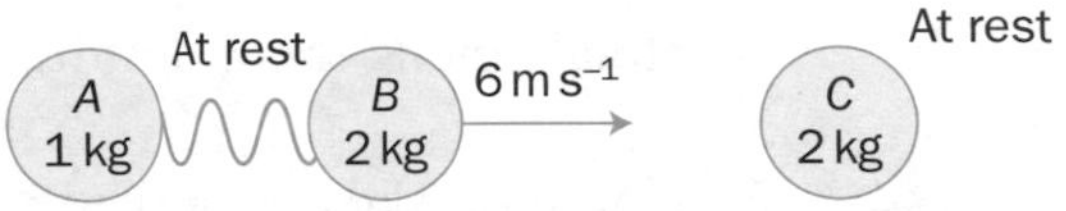

a the common speed of A and B after the string becomes taut

b the impulse suffered by C when B collides with it

c the new common speed of A and B after the string becomes taut for the second time, and confirm that there are no more collisions.

5 Three identical particles A, B and C lie at rest in that order in a straight line. A is projected at speed u towards B. The coefficient of restitution between all the particles is e.

a Find, in terms of u and e, the velocities of the particles after two collisions have taken place.

b Show that there will be at least one more collision unless $e = 1$.

M2

1 A particle A, of mass 3 kg, rests on a smooth horizontal surface. A second particle B, of mass 1 kg, is propelled towards A at $5\,\text{m}\,\text{s}^{-1}$ from a point P 10 m from A. B rebounds from A and returns to P. The coefficient of restitution is 0.6. Calculate

a the length of time between B's leaving P and its return

b the distance between P and the particle A at the instant that B returns to P.

2 A particle A of mass 4 kg and moving at $6\,\text{m}\,\text{s}^{-1}$ collides with a second particle B, of mass 1 kg and moving in the same direction along the same straight line at $2\,\text{m}\,\text{s}^{-1}$. After the collision, A moves in the same direction with its speed reduced to $5\,\text{m}\,\text{s}^{-1}$.

a Calculate the new velocity of B.

b Calculate the coefficient of restitution between the particles.

3 Two particles, P and Q, move towards each other on a smooth horizontal surface. They have masses of 2 kg and 5 kg and speeds of $1\,\text{m}\,\text{s}^{-1}$ and $3\,\text{m}\,\text{s}^{-1}$ respectively. The coefficient of restitution between them is 0.75. Find

a the velocities of the particles after they collide

b the magnitude of the impulse acting on Q

c the total kinetic energy lost in the collision.

4 A block of mass 2 kg is propelled across a rough horizontal surface at $8\,\text{m}\,\text{s}^{-1}$ directly towards a vertical wall 10 m away. The coefficient of friction between the block and the surface is $\frac{1}{7}$ and the coefficient of restitution between the block and the wall is $\frac{2}{3}$.

a Calculate the speed of the block immediately before and after striking the wall.

b Find the distance from the wall at which the block comes to rest.

5 A ball of mass m is thrown vertically downwards so that it strikes horizontal ground travelling at $10\,\text{m}\,\text{s}^{-1}$. The coefficient of restitution between the ball and the ground is 0.7. Calculate

a the height to which the ball bounces after its first contact with the ground

b the height to which it rises on the second bounce

c the total distance travelled by the ball between the first bounce and when it comes to rest.

6 The diagram shows two spheres A and B, of mass 1 kg and 3 kg, which are initially at rest on a smooth horizontal surface. AB is perpendicular to a vertical wall. The spheres are joined by a light inextensible string which is initially slack. The coefficient of restitution between the spheres is 0.5, and between B and the wall is 0.6. B is propelled towards the wall at $4\,\text{m s}^{-1}$.

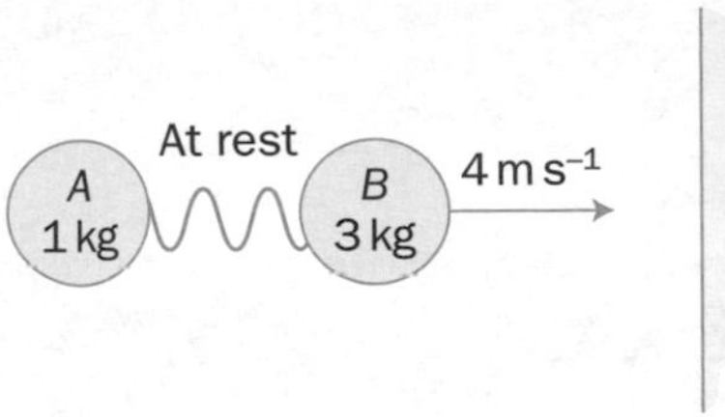

a Assuming that B strikes the wall before the string becomes taut, find

i the speed with which B leaves the wall

ii the velocities of A and B immediately after the impact between the spheres

iii the common speed of the spheres after the string becomes taut.

b Find how the situation will be different if the string becomes taut before B strikes the wall, and find the loss of kinetic energy during the whole process if this is the case.

7 Spheres A, B and C, having masses of 1 kg, 3 kg and 3 kg respectively, lie at rest in a straight line on a smooth horizontal surface with B between A and C. Sphere A is propelled towards B with speed V. The coefficient of restitution between A and B is $\frac{1}{3}$ and between B and C is $\frac{1}{2}$.

a Show that A is brought to rest in the first collision, and find the speed of B immediately after this.

b Find the velocities of B and C after they collide. Hence show that

i there will be a total of two collisions

ii the total loss of kinetic energy is $\frac{5V^2}{48}$.

8 Points A and B are in a vertical line, with B 5 m above A. A particle of mass m is projected upwards from A at $14\,\text{m s}^{-1}$. At the same moment an identical particle is projected upwards from B at $7\,\text{m s}^{-1}$. The coefficient of restitution between the particles is 0.5.

a Show that the particles collide at a point 7.5 m above A.

b Find the speed of each particle when it returns to its starting point.

M2

4 Exit

Summary

Refer to

- Impulse $\mathbf{I} = \mathbf{F}t$ N s
  - Momentum $= m\mathbf{v}$ N s
  - Impulse = change of momentum:
    $\mathbf{I} = m\mathbf{v} - m\mathbf{u}$
- If no external force acts on a system, the total momentum of the system remains constant. This is the principle of conservation of momentum. 4.1
- Newton's experimental law of restitution

  $$\frac{\text{separation speed}}{\text{approach speed}} = e \text{ (a constant)}$$

  - If $e = 0$ the impact is perfectly inelastic. No rebounding occurs.
  - If $e = 1$ the impact is perfectly elastic. 4.2
- Kinetic energy is not conserved unless a collision is perfectly elastic ($e = 1$). 4.2
- For collisions between objects which are free to move momentum is conserved and Newton's law of restitution applies. 4.2
- For collisions with fixed objects momentum is not conserved but Newton's law of restitution applies. 4.3, 4.4

Links

Momentum and restitution are key tools in the study of vehicle collisions, both as part of accident investigation after an accident, or as part of the design and testing of new vehicles.

Modern vehicles are fitted with safety features such as airbags and bumpers which act to reduce the effect of a collision on the passenger.

Engineers can use principles such as the conservation of momentum to understand the forces which act on a passenger following a collision and to attempt to reduce the risk of injury as much as possible.

5 Statics of rigid bodies

This chapter will show you how to
- understand the conditions necessary for a rigid body to be in equilibrium
- calculate unknown forces acting on a rigid body in equilibrium.

Before you start

You should know how to:	Check in:
1 Resolve a force into components.	1 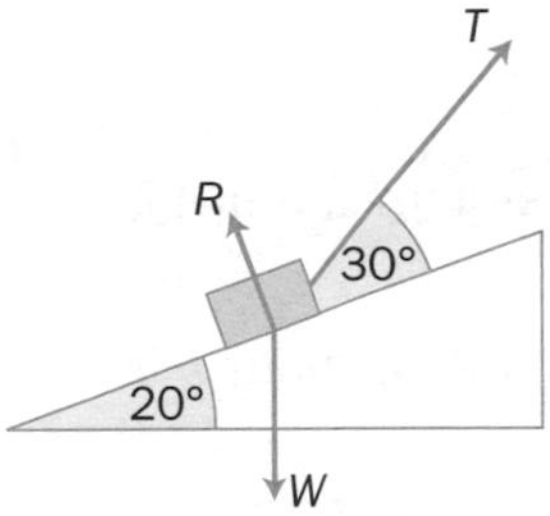 The diagram shows a block on a smooth inclined plane. Write down a the total force acting parallel to the plane b the total force perpendicular to the plane.
2 Calculate the moment of a force about a chosen point.	2 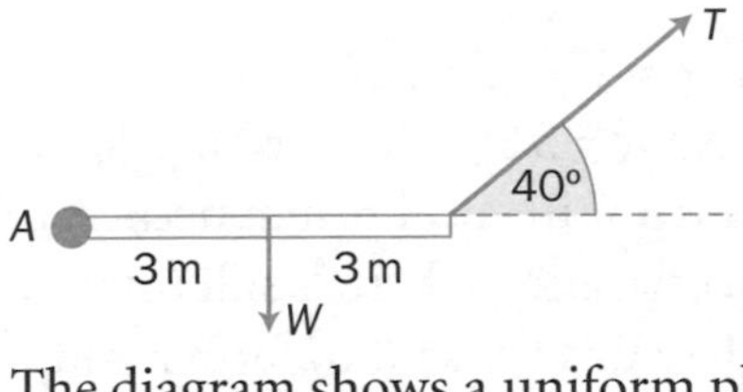 The diagram shows a uniform plank, 6 m long, hinged at end A and supported at the other end by a string. Find the total moment of W and T about the hinge.
3 Use the standard model of friction.	3 A block rests on a rough surface under the action of several forces, including friction. The normal reaction force is R and the friction force is F. The coefficient of friction is 0.3. What can you say about F and R if a you know the block is on the point of slipping b you don't know whether the block is about to slip?

M2

5.1 Conditions for equilibrium

In Mechanics 1 you learned about parallel forces and the conditions under which they are in equilibrium.

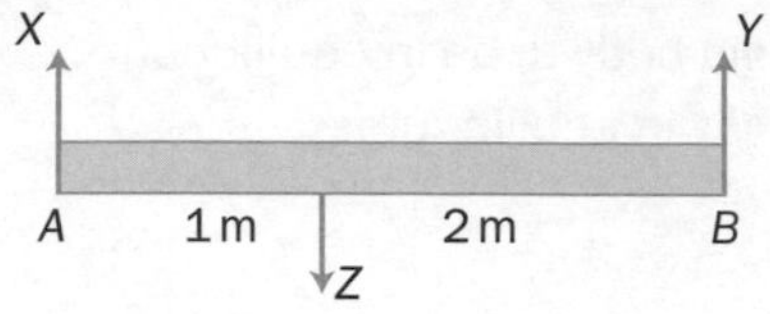

This diagram shows a rod of length 3 m acted on by three forces X, Y and Z which are perpendicular to the rod.

The resultant of the forces is

$R = X + Y - Z$

The moment of the system about A is

$M_A = 3Y - Z$

The anti-clockwise direction is taken to be positive, as is the convention.

The moment of the system about B is

$M_B = 2Z - 3X$

Consider the following cases.

a $X = 5\,\text{N}$, $Y = 1\,\text{N}$, $Z = 3\,\text{N}$
These give $R = 3\,\text{N}$, $M_A = 0\,\text{N m}$, $M_B = -9\,\text{N m}$
The rod is not in equilibrium because it has linear acceleration in the direction of X and Y, as well as rotational acceleration.

b $X = 3\,\text{N}$, $Y = 2\,\text{N}$, $Z = 5\,\text{N}$
These give $R = 0\,\text{N}$, $M_A = 1\,\text{N m}$, $M_B = 1\,\text{N m}$
The rod is not in equilibrium because, although it has no linear acceleration, there is still rotational acceleration (forces like this are said to form a couple).

c $X = 4\,\text{N}$, $Y = 2\,\text{N}$, $Z = 6\,\text{N}$
These give $R = 0\,\text{N}$, $M_A = 0\,\text{N m}$, $M_B = 0\,\text{N m}$
The rod is now in equilibrium because there is no linear acceleration and no rotational acceleration.

For the forces to be in equilibrium the resultant force must be zero and the total moment about any point must be zero.

The conditions under which parallel forces are in equilibrium also apply to systems of non-parallel forces. To be sure that the resultant force is zero you need to check the total of the components in each of two directions.

You will only be expected to deal with a system of forces all acting in the same plane – that is, coplanar forces.

Conditions for equilibrium

A system of coplanar forces is in equilibrium if

- the resultant force in each of two directions is zero
- the total moment about any chosen point is zero.

So for a system in equilibrium you can obtain three equations, by resolving in each of two directions and taking moments about a point.

You could go on and resolve in other directions or take moments about other points, but the equations obtained would contain no extra information. They would just be combinations of the original three equations.

EXAMPLE 1

A ladder AB of mass 20 kg rests on smooth horizontal ground and leans against a smooth vertical wall. The inclination of the ladder to the horizontal is 60°. The ladder is kept in position by a horizontal force P N applied to the bottom of the ladder. Find the value of P and the reactions at the wall and the ground.

Suppose the ladder has length $2a$. There are just normal reactions R and S at the ground and the wall because the contacts are smooth. The system is in equilibrium.

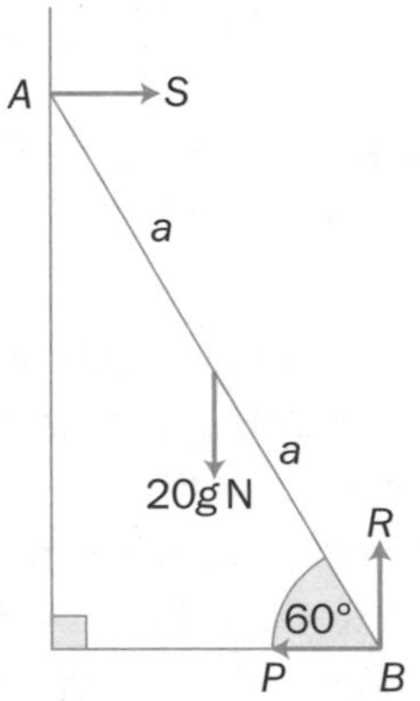

Resolve vertically:

$$R - 20g = 0 \quad \text{and so} \quad R = 196\,\text{N}$$

Resolve horizontally:

$$S - P = 0 \qquad [1]$$

Take moments about B:

$$20ag\cos 60° - 2aS\sin 60° = 0 \qquad [2]$$

From [2] $\quad S = \left(\frac{10g}{\sqrt{3}}\right)\text{N} = 56.6\,\text{N}$

From [1] $\quad P = 56.6\,\text{N}$

Example 1 uses the most common strategy for solving equilibrium problems:

- Resolve in two (usually perpendicular) directions. In each case, equate the total of the components to zero.
- Take moments about one chosen point. Equate the total moment to zero.

Take care when choosing the directions for resolving and the point about which to take moments. In Example 1 you could resolve parallel and perpendicular to the ladder and take moments about the middle of the ladder, but all three equations would contain P, R and S. They would give the correct values, but the algebra would be more tedious.

There are two other strategies which can sometimes prove useful:

Alternative strategies

1. Resolve in one direction and take moments about two points.
2. Take moments about three points.

In 1, if you take moments about P and Q you must not resolve perpendicular to PQ. In 2, the three points must not be collinear.

M2

EXAMPLE 2

A uniform rod AB, of length 2 m and mass 5 kg, rests with A on smooth horizontal ground and B on a rough peg 1 m above the ground. Find the reaction at A and the normal reaction and friction forces at B.

From the dimensions given, the angle at A is 30°.

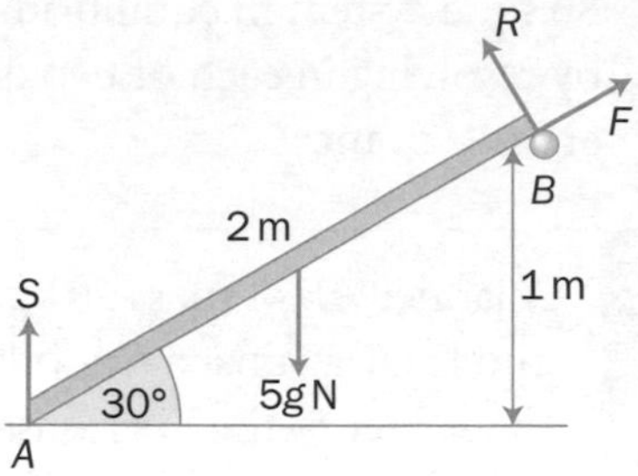

Take moments about B:

$5g\cos 30° \times 1 - S\cos 30° \times 2 = 0$

giving $S = (2.5g)\text{ N} = 24.5\text{ N}$

Take moments about A:

$R \times 2 - 5g\cos 30° \times 1 = 0$

giving $R = (2.5g\cos 30°)\text{ N} = 21.2\text{ N}$

Resolve along AB:

$F + S\sin 30° - 5g\sin 30° = 0$

giving $F = (2.5g\sin 30°)\text{ N} = 12.25\text{ N}$

Example 3 gives a rare instance where taking moments about three points gives the most elegant solution.

EXAMPLE 3

A fixed smooth cylinder, radius a and centre O, rests on a smooth horizontal surface with its axis horizontal. A rod AB of weight W rests with A on the horizontal surface and B on the cylinder such that AB is inclined at 60° to the horizontal and is a tangent to the cylinder. The rod is held in place by a light string, AP, attached to the cylinder at P so that APO is a straight line. Find the tension in the string and the reactions at A and B.

C is the point where OB meets the vertical through A, and E is the point on the horizontal surface below the centre of the rod, as shown.

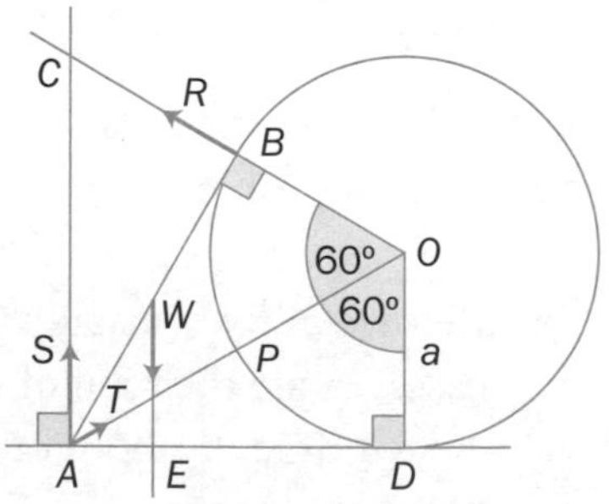

Triangle AOC is equilateral, so CP is perpendicular to AO and

$AB = AD = CP = a\sqrt{3}$

Also $AE = \frac{1}{4}AD = \frac{1}{4}a\sqrt{3}$

Take moments about A:

$Ra\sqrt{3} - \frac{1}{4}Wa\sqrt{3} = 0$ which gives $R = \frac{1}{4}W$

Take moments about O:

$\frac{3}{4}Wa\sqrt{3} - Sa\sqrt{3} = 0$ which gives $S = \frac{3}{4}W$

Take moments about C:

$Ta\sqrt{3} - \frac{1}{4}Wa\sqrt{3} = 0$ which gives $T = \frac{1}{4}W$

EXAMPLE 4

A ladder AB, of length $4a$ and weight W, rests with A on rough horizontal ground and B against a rough vertical wall. The coefficient of friction in both contacts is 0.5. The ladder is at 60° to the horizontal. Investigate whether a person of weight $9W$ can climb all the way to the top.

Suppose the ladder slips when the person reaches P, where $AP = x$, as shown.
When the ladder is in limiting equilibrium (on the point of slipping), you have

$$F_1 = 0.5R_1 \quad [1]$$
$$F_2 = 0.5R_2 \quad [2]$$

Resolve vertically and horizontally:

$$R_1 + F_2 = 10W \quad [3]$$
$$F_1 = R_2 \quad [4]$$

Take moments about A:

$$Wa + \frac{9Wx}{2} = 2F_2\ a + 2R_2\ a\sqrt{3} \quad [5]$$

Solve equations [1] to [4]:

$$R_1 = 8W,\ F_1 = 4W,\ R_2 = 4W,\ F_2 = 2W$$

Substitute into [5] and cancel through by W:

$$a + \frac{9x}{2} = 4a + 8a\sqrt{3} \quad \text{so} \quad x = \frac{2a(3 + 8\sqrt{3})}{9} = 3.75a$$

So the ladder will slip before the person gets to the top.

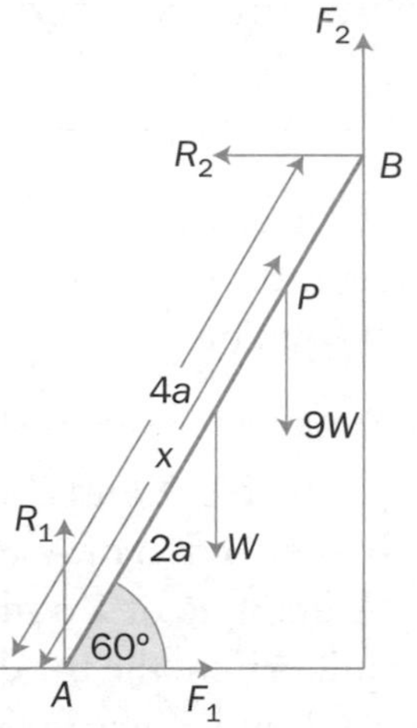

This solution models the person as a particle. In practice, the person is significantly large in relation to the size of the ladder and tends to exert force at more than one point.
It also begs the question as to what it means for a person to reach the top of a ladder.

M2

Some problems require you to find the reaction in a hinge or a joint. This is usually best achieved by finding the horizontal and vertical components of the reaction.

EXAMPLE 5

A rod AB of length a and weight W is hinged to a vertical wall at A and is held at an angle of 30° above the horizontal by a light string BC, also of length a, which is fixed to the wall at C, a distance a vertically above A. Find the reaction in the hinge at A.

Let the reaction have horizontal and vertical components, X and Y, as shown.

Take moments about C:

$$Xa - \tfrac{1}{2}Wa\sin 60° = 0$$

giving $$X = \tfrac{1}{4}W\sqrt{3}$$

Take moments about B:

$$\tfrac{1}{2}Wa\sin 60° + Xa\cos 60° - Ya\sin 60° = 0$$

giving $Y\sqrt{3} = X + \tfrac{1}{2}W\sqrt{3}$ and so $Y = \tfrac{3}{4}W$

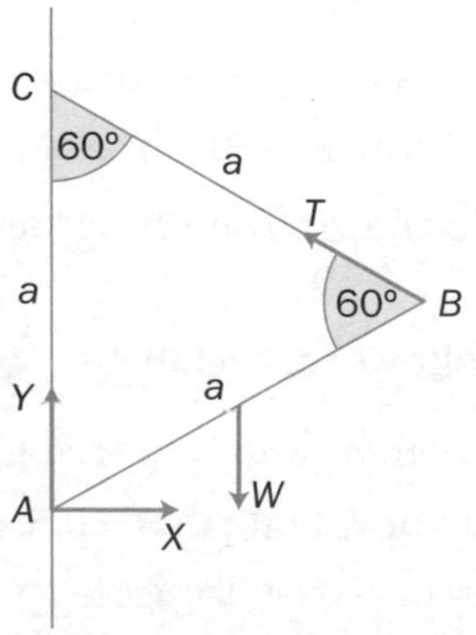

Example 5 continues on the next page.

EXAMPLE 5 (CONT.)

Combine X and Y to find the magnitude of the reaction R:

$$R = \sqrt{X^2 + Y^2} = \tfrac{1}{2}W\sqrt{3}$$

The reaction makes an angle θ to the horizontal, where

$$\tan\theta = \frac{Y}{X} = \sqrt{3} \quad \text{and so} \quad \theta = 60^\circ$$

You have already met problems involving the equilibrium of three concurrent forces. In fact in all problems with three non-parallel forces in equilibrium, the forces are concurrent.

Concurrent forces all act through the same point.

Suppose that X, Y and Z are three non-parallel forces in equilibrium.
The lines of action of X and Y meet at some point A.
The resultant, R, of X and Y acts through A.
For equilibrium, the third force, Z, must exactly oppose R.
So $Z = -R$, and Z acts through A.
The three forces X, Y and Z are therefore concurrent.

When a system of three non-parallel forces is in equilibrium, the lines of action of the forces must all pass through a single point.

You can sometimes use this fact to help solve problems.

EXAMPLE 6

A uniform rod AB of weight W rests with end B against a smooth vertical wall. End A is hinged to horizontal ground. The rod is inclined at 45° to the horizontal. Find the magnitude and direction of S, the reaction in the hinge, and the magnitude of the reaction R between the rod and the wall.

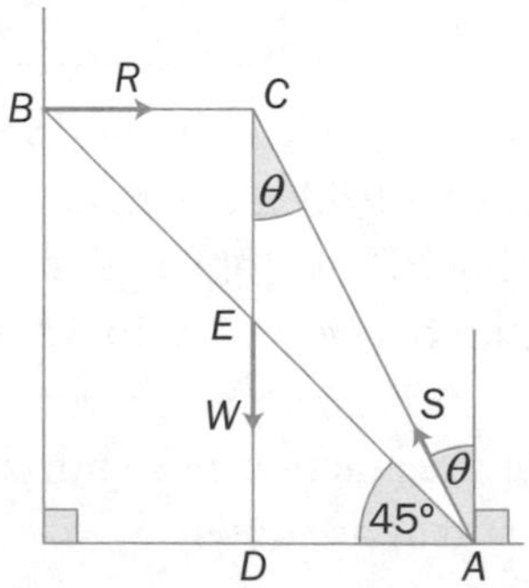

S makes an angle θ with the vertical, as shown.

The lines of action of R and W meet at C. There are only three forces, so the line of action of S must also pass through C.

BCE and ADE are congruent right-angled isosceles triangles, so $CD = 2AD$.

It follows that $\tan\theta = \dfrac{AD}{CD} = 0.5$ giving $\theta = 26.6^\circ$

You can now draw a triangle of forces, as shown.

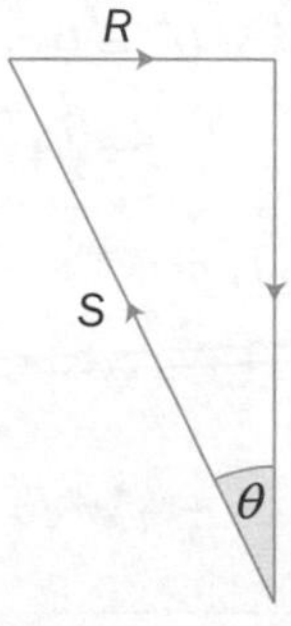

You know that $\tan\theta = 0.5$, and hence $R = \tfrac{1}{2}W$

Apply Pythagoras' theorem:

$$S = \sqrt{W^2 + R^2} = \frac{W\sqrt{5}}{2}$$

Exercise 5.1

1 A uniform rod AB of length 3 metres and mass 15 kg is hinged at A. A light string is attached to B and holds the rod in equilibrium at an angle of 60° to the upward vertical through A. Find the tension in the string when

a the string is at right angles to AB

b the string is vertical

c the string is horizontal.

2 The diagram shows a rod AB of length 2 m and mass 4 kg. The rod is freely hinged at A to a point on a vertical wall and makes an angle of 50° with the wall. The rod is held in equilibrium by a horizontal force of T N applied at B.

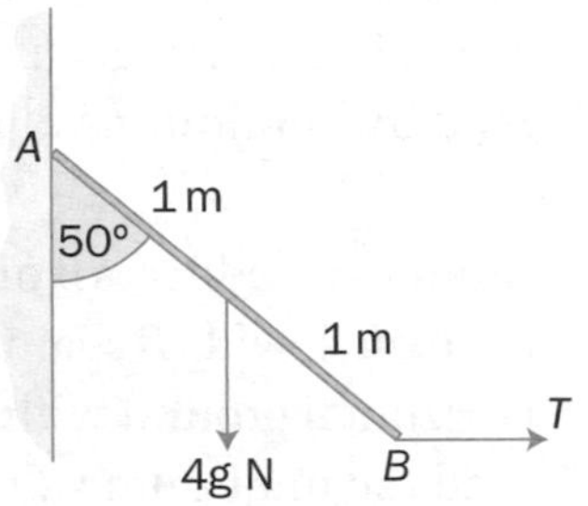

a Calculate the value of T.

b Find the magnitude and direction of the reaction at the hinge.

3 A non-uniform rod AB of length 5 m and mass 6 kg is in equilibrium suspended in a horizontal position on two light strings. The strings make angles of 50° and 30° with the horizontal. The centre of mass of the rod is at C, where $AC = x$.

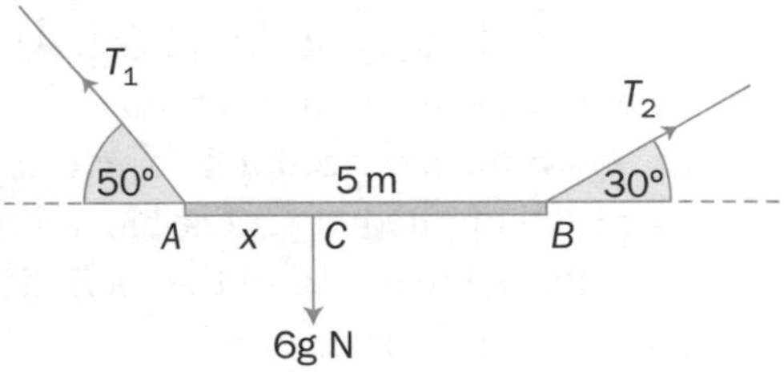

a Calculate T_1 and T_2, the tensions in the strings.

b Find the value of x.

4 The diagram shows a horizontal uniform rod AB of length $2a$ and weight W. A light string is attached to A and B and passes through a smooth ring at C, vertically above A, so that angle $A\hat{B}C$ is 30°. A horizontal force P is applied to A.

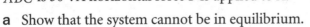

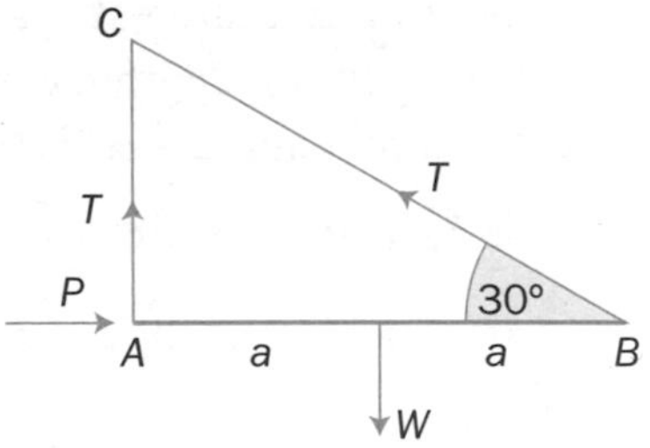

a Show that the system cannot be in equilibrium.

A weight W is now attached at a point D on AB so that equilibrium is maintained.

b Find the distance AD and the force P.

5 A uniform rod AB of mass 10 kg and length 2 m rests with A on smooth horizontal ground and B on a smooth peg 1 m above the ground. The rod is held in position by a horizontal force of P N at A. Find the value of P and the magnitude of the reactions at A and B.

6 A uniform ladder AB of mass 20 kg and length 3 m rests with B against a smooth wall and A on smooth horizontal ground. A is attached by means of a light inextensible string, 1 m long, to the base of the wall.

a Find the tension in the string.

b If the breaking strain of the string is 250 N, find how far up the ladder a man of mass 80 kg can safely ascend.

7 A uniform ladder AB of weight W rests at an angle α to the horizontal with B against a smooth vertical wall and A on rough horizontal ground with coefficient of friction 0.25. Find the minimum value of α if the ladder does not slip.

8 A uniform ladder AB of weight W rests at an angle α to the horizontal with B against a rough vertical wall and A on rough horizontal ground with coefficient of friction 0.25 at each contact. Find the minimum value of α if the ladder does not slip.

9 The diagram shows a cross-section $ABCD$ of a uniform rectangular block of mass 20 kg. AB is 0.75 m and BC is 1 m. The block rests with A on rough horizontal ground and AB at 20° to the horizontal. It is held in place by a horizontal force PN applied at C. The block is on the point of slipping. Find the value of P and the coefficient of friction between the block and the ground.

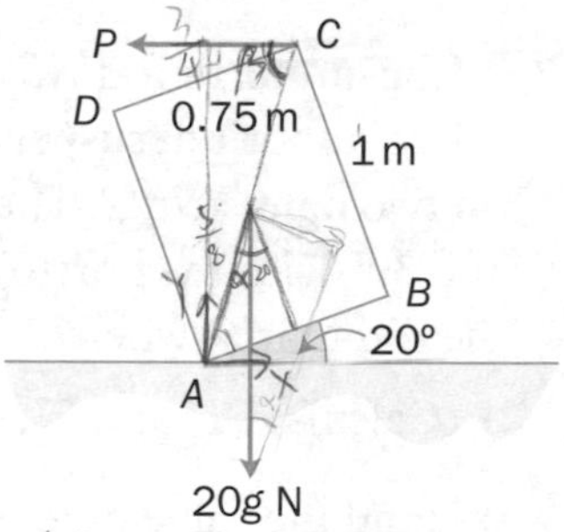

10 A uniform rod AB of length $2a$ and weight W is hinged to a horizontal ceiling at A and is suspended by a light inextensible string BC of length a connecting B to a point C on the ceiling such that angle ABC is 90°. Show that the tension in the string is $\frac{W}{\sqrt{5}}$, and find the horizontal and vertical components of the reaction of the hinge on the rod.

11 The diagram shows a cross-section of a uniform horizontal shelf hinged to a vertical wall at A. $AB = 2a$ and the shelf has weight W. The shelf is supported by a light string CD connecting a point D on the shelf to a point C on the wall. $AD = x$ and $AC = a$. The breaking strain of the string is $4W$.

a Find the minimum value of x.

b For this value of x find the magnitude and direction of the reaction at the hinge.

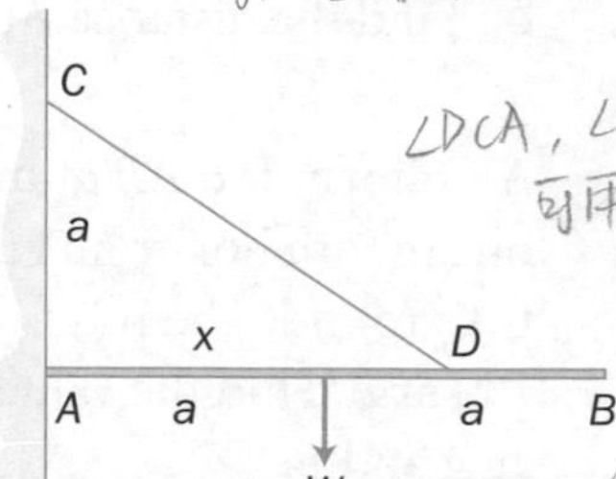

12 A uniform ladder AB rests in equilibrium with B against a smooth vertical wall and A on rough horizontal ground. The coefficient of friction between the ladder and the ground is μ. The ladder makes an angle of θ to the horizontal and is on the point of slipping. Show that

$$\tan\theta = \frac{1}{2\mu}$$

13 The diagram shows a sign outside a shop. It comprises a uniform lamina $ABCD$ in the shape of a square attached to a triangle. The mass of the lamina is 6 kg. The sign is mounted in a vertical plane with AD horizontal. It is smoothly hinged to the wall at A and is held in equilibrium by a horizontal string BE attached to the wall at E.

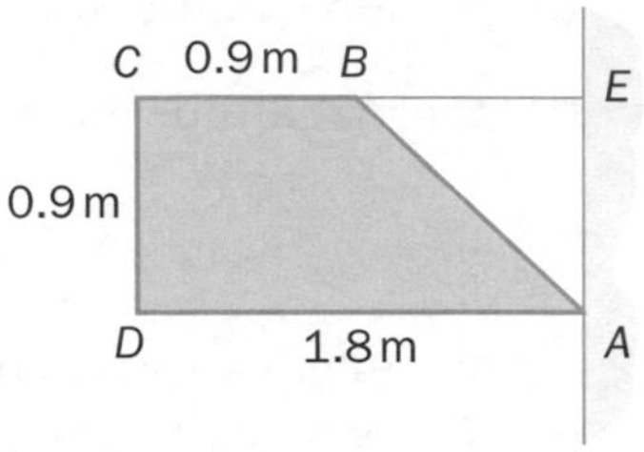

a Calculate the tension in the string.

b Find the magnitude and direction of the reaction at the hinge.

14 A uniform ladder AB has length 8 m and mass 20 kg. It leans against a rough wall with end A on rough horizontal ground. The coefficient of friction at both contacts is 0.3 and the ladder makes an angle of 65° with the horizontal. A woman of mass 50 kg starts to climb the ladder.

a How far can she ascend before the ladder is on the point of slipping?

Her daughter, of mass m kg, stands on the ladder at A. This enables the woman to climb right to the top of the ladder.

b Find the least value of m.

15 The diagram shows a uniform rod AB of weight W and length $2a$. The rod rests with A on rough horizontal ground and leans against a rough fixed prism of semicircular cross-section of radius a. The coefficient of friction at both contacts is μ. When friction is limiting the rod makes an angle θ with the horizontal. Show that

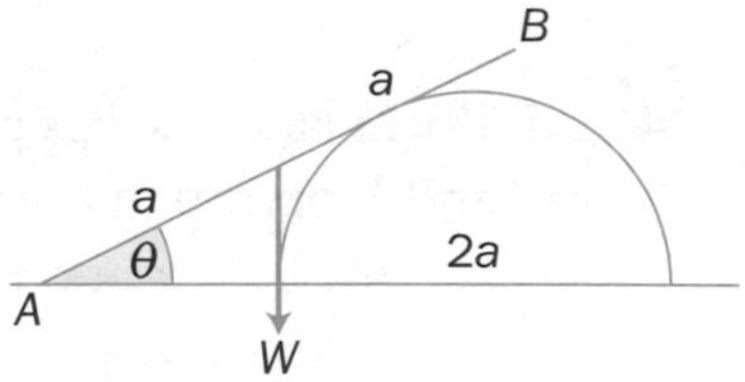

$$\sin\theta = \sqrt{\frac{\mu}{1+\mu^2}}$$

16 The diagram shows a uniform ladder leaning against a smooth vertical wall. The ladder stands on rough ground (coefficient of friction μ) which slopes away from the wall at an angle of α to the horizontal. The ladder makes an angle of θ to the horizontal. If the ladder is in limiting equilibrium, show that

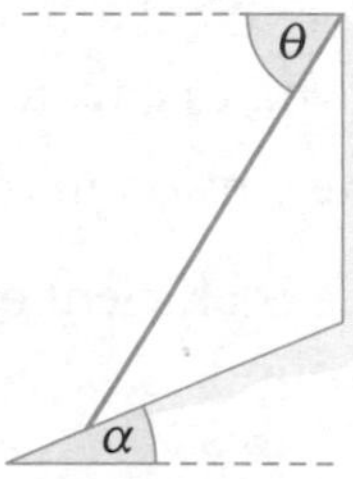

$$\tan\theta = \frac{1+\mu\tan\alpha}{2(\mu-\tan\alpha)}$$

M2

1 A uniform beam AB of length 2 m and weight 40 N is smoothly hinged to a vertical wall at B. It is held in a horizontal position by a rod CD of length 1 m attached to the beam at C, where $CB = 0.5$ m, and to the wall below B. Calculate

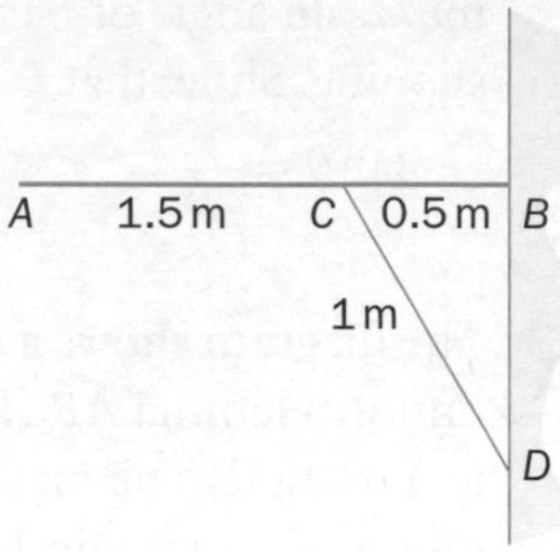

a the force in the rod

b the magnitude and direction of the reaction at the hinge.

2 A non-uniform rod AB of length 5 m and mass 40 kg is suspended in a horizontal position by light strings attached to A and B. The strings make angles of 20° and 60° with the horizontal, as shown. Find

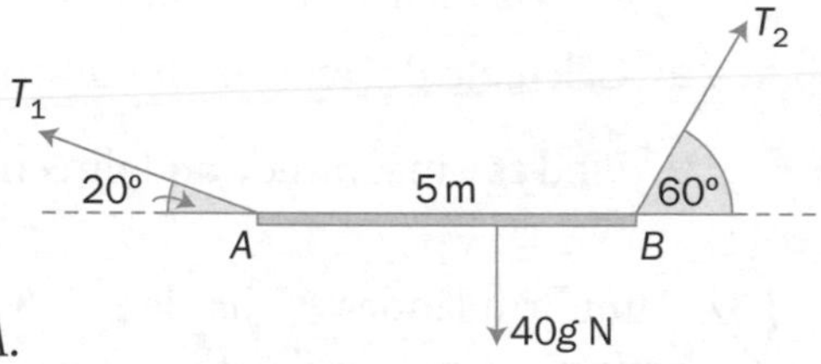

a the tensions, T_1 and T_2, in the strings

b the distance of the centre of mass of the rod from A.

3 A uniform ladder AB of length 4 m and mass 10 kg rests in equilibrium with B against a smooth vertical wall and A on a smooth horizontal surface 2 m from the base of the wall. The ladder is held in equilibrium by a rope fastened to the base of the wall. Find the tension in the rope if

a it is attached to the ladder at A

b it is attached to the ladder so that the rope is perpendicular to the ladder.

4 A uniform ladder AB is placed with B against a smooth vertical wall and A on rough horizontal ground. Find the coefficient of friction between the ladder and the ground if the ladder is in limiting equilibrium when it makes an angle of 60° with the horizontal.

5 A uniform beam AB of length 4 m and mass 30 kg rests in limiting equilibrium at 40° to the horizontal, with A on rough horizontal ground and B supported by a rope which is perpendicular to the beam and in the same vertical plane, as shown. Calculate

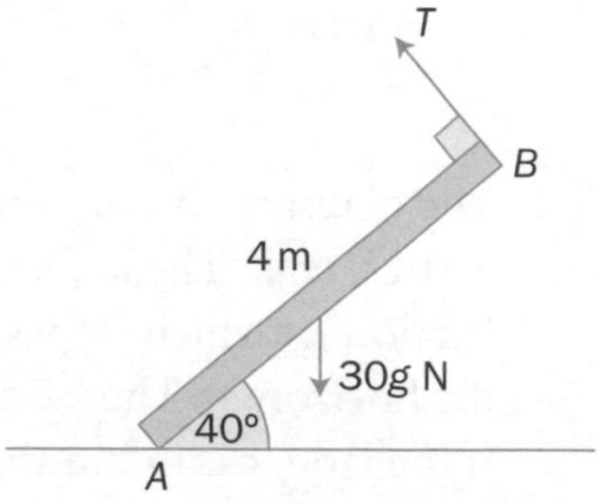

a the tension in the rope

b the coefficient of friction between the beam and the ground.

6 The diagram shows a thin metal rod of weight 20 N bent to form a semicircular arc, with diameter $AB = 2$ m. The rod is smoothly hinged to a fixed support at A and is held in equilibrium with AB vertical by a horizontal string attached to B. Calculate

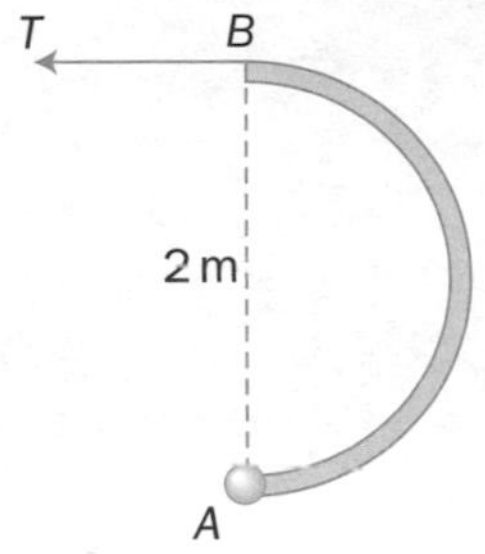

a the tension in the string

b the magnitude and direction of the reaction at the hinge.

7 A uniform rod of length $4a$ and weight W rests in limiting equilibrium supported by two rough pegs A and B. A is above the end of the rod and B is underneath the rod a distance a along it, as shown. The rod is at 30° to the horizontal. The coefficient of friction at each of the two pegs is μ. Show that

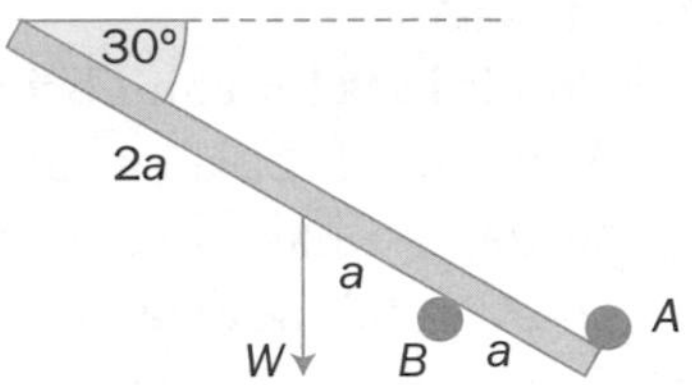

$$\mu = \frac{1}{3\sqrt{3}}$$

8 A uniform ladder AB of length $2a$ and weight W rests with B against a smooth vertical wall and A on rough horizontal ground. The ladder makes an angle of 60° with the horizontal. A child of weight W can just climb to the top of the ladder without causing it to slip. Find how far up the ladder a woman of weight $4W$ could safely climb.

9 A uniform rod AB of length 2 m and weight 12 N is placed with B against a rough vertical wall and A on rough horizontal ground. The rod makes an angle of 50° with the horizontal. The coefficient of friction at both contacts is 0.2.

a Show that with no additional forces applied the rod will slip.

b Find the minimum horizontal force which must be applied at A to prevent the rod from slipping.

c Find the maximum horizontal force which could be applied at A without causing the rod to slip.

10 The diagram shows a uniform rod AB of weight $2W$ and length $2a$ which is smoothly hinged to a fixed support at A. A light inextensible string is attached to a smooth ring which is free to slide on the rod. The string passes over a smooth pulley at C, a distance $3a$ vertically above A, and carries on its end a particle of weight W. The rod makes an angle θ with the vertical.

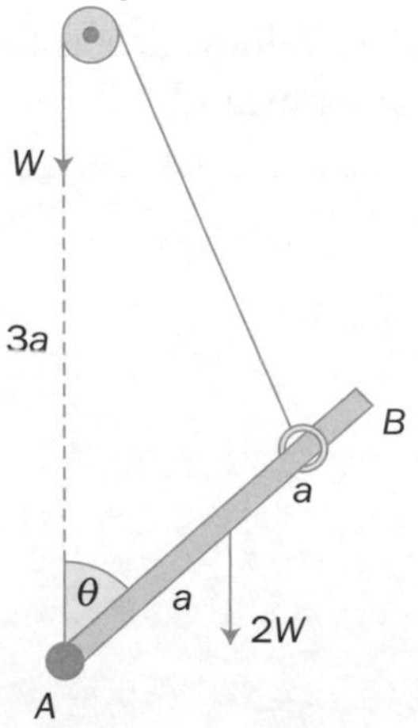

a Explain why the string will be perpendicular to the rod when the system is in equilibrium.

b Show that $\tan\theta = \frac{3}{2}$

c Find the magnitude of the reaction at the hinge.

M2

5 Exit

Summary

Refer to

- **Conditions for equilibrium**

  A system of coplanar forces is in equilibrium if
  - the resultant force in each of two non-parallel directions is zero
  - the total moment about any chosen point is zero.
- There are three strategies for obtaining equations leading to the solution of a problem.
  - Resolve in two (usually perpendicular) directions and take moments about one chosen point. Equate the totals in each case to zero.
  - Resolve in one direction and take moments about two points.
  - Take moments about three points (which are not collinear).
- When a system of three non-parallel forces is in equilibrium, the lines of action of the forces must all pass through a single point. 5.1

Links

Architects and structural engineers use knowledge of statics extensively when designing stationary structures such as bridges, buildings and tunnels.

There are several different basic bridge designs. Engineers must choose the most appropriate design for a specific bridge, and some times they must combine aspects from the different basic designs within one bridge.

By calculating the forces acting on a bridge, the engineers are able to ensure that it does not collapse and to determine safety values such as the maximum load that the bridge can withstand.

Revision exercise 2

1 A car of mass 1000 kg is moving along a straight horizontal road with a constant acceleration of $f\,\text{m s}^{-2}$. The resistance to motion is modelled as a constant force of magnitude 1200 N. When the car is travelling at $12\,\text{m s}^{-1}$, the power generated by the engine of the car is 24 kW.

a Calculate the value of f.

When the car is travelling at $14\,\text{m s}^{-1}$, the engine is switched off and the car comes to rest, without braking, in a distance of d metres. Assuming the same model for resistance

b use the work-energy principle to calculate the value of d.

c Give a reason why the model used for the resistance to motion may not be realistic. [(c) Edexcel Limited 2003]

2 A girl and her bicycle have a combined mass of 64 kg. She cycles up a straight stretch of road which is inclined at an angle α to the horizontal, where $\sin\alpha = \frac{1}{14}$. She cycles at a constant speed of $5\,\text{m s}^{-1}$. When she is cycling at this speed, the resistance to motion from non-gravitational forces has magnitude 20 N.

a Find the rate at which the cyclist is working.

She now turns round and comes down the same road. Her initial speed is $5\,\text{m s}^{-1}$, and the resistance to motion is modelled as remaining constant with magnitude 20 N. She free-wheels down the road for a distance of 80 m. Using this model

b find the speed of the cyclist when she has travelled a distance of 80 m.

The cyclist again moves down the same road, but this time she pedals down the road. The resistance is now modelled as having magnitude proportional to the speed of the cyclist. Her initial speed is again $5\,\text{m s}^{-1}$ when the resistance to motion has magnitude 20 N.

c Find the magnitude of the resistance to motion when the speed of the cyclist is $8\,\text{m s}^{-1}$.

The cyclist works at a constant rate of 200 W.

d Find the magnitude of her acceleration when her speed is $8\,\text{m s}^{-1}$. [(c) Edexcel Limited 2003]

3 A particle of mass 0.8 kg is moving in a straight line on a rough horizontal plane. The speed of the particle is reduced from $15\,\text{m s}^{-1}$ to $10\,\text{m s}^{-1}$ as the particle moves 20 m. Assuming that the only resistance to motion is the friction between the particle and the plane, find

a the work done by friction in reducing the speed of the particle from $15\,\text{m s}^{-1}$ to $10\,\text{m s}^{-1}$

b the coefficient of friction between the particle and the plane. [(c) Edexcel Limited 2007]

4 Two particles A and B, of mass m and $2m$ respectively, are attached to the ends of a light inextensible string. The particle A lies on a rough plane inclined at an angle α to the horizontal, where $\tan \alpha = \frac{3}{4}$. The string passes over a small light smooth pulley P fixed at the top of the plane. The particle B hangs freely below P, as shown in the diagram. The particles are released from rest with the string taut and the section of the string from A to P parallel to a line of greatest slope of the plane. The coefficient of friction between A and the plane is $\frac{5}{8}$. When each particle has moved a distance h, B has not reached the ground and A has not reached P.

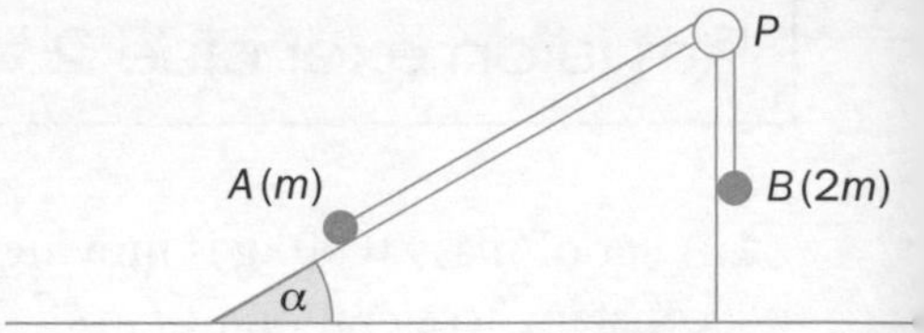

a Find an expression for the potential energy lost by the system when each particle has moved a distance h.

When each particle has moved a distance h they are moving with speed v. Using the work-energy principle

b find an expression for v^2, giving your answer in the form kgh, where k is a number.

[(c) Edexcel Limited 2007]

5 A package of mass 3.5 kg is sliding down a ramp. The package is modelled as a particle and the ramp as a rough plane inclined at an angle of 20° to the horizontal. The package slides down a line of greatest slope of the plane from a point A to a point B, where $AB = 14$ m. At A the package has speed 12 m s^{-1} and at B the package has speed 8 m s^{-1}, as shown. Find

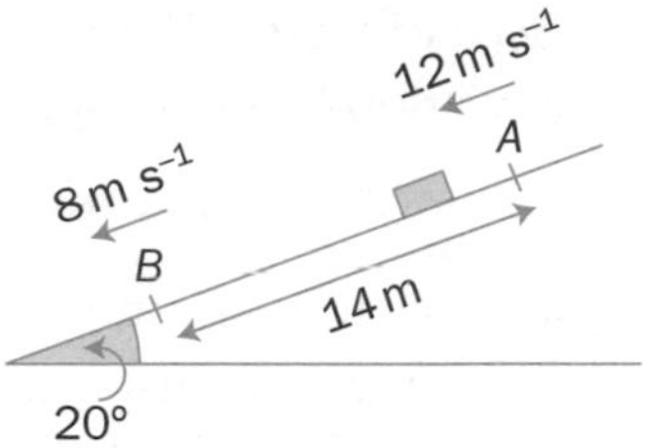

a the total energy lost by the package in travelling from A to B

b the coefficient of friction between the package and the ramp.

[(c) Edexcel Limited 2008]

6 A car of mass 1000 kg is moving at a constant speed of 16 m s^{-1} up a straight road inclined at an angle θ to the horizontal. The rate of working of the engine of the car is 20 kW and the resistance to motion from non-gravitational forces is modelled as a constant force of magnitude 550 N.

a Show that $\sin \theta = \frac{1}{14}$

When the car is travelling up the road at 16 m s^{-1}, the engine is switched off. The car comes to rest, without braking, having moved a distance y metres from the point where the engine was switched off. The resistance to motion from non-gravitational forces is again modelled as a constant force of magnitude 550 N.

b Find the value of y.

[(c) Edexcel Limited 2008]

M2

7 A smooth sphere P of mass $2m$ is moving in a straight line with speed u on a smooth horizontal table. Another smooth sphere Q of mass m is at rest on the table. The sphere P collides directly with Q. The coefficient of restitution between P and Q is $\frac{1}{3}$. The spheres are modelled as particles.

a Show that, immediately after the collision, the speeds of P and Q are $\frac{5}{9}u$ and $\frac{8}{9}u$ respectively.

After the collision, Q strikes a fixed vertical wall which is perpendicular to the direction of motion of P and Q. The coefficient of restitution between Q and the wall is e. When P and Q collide again P is brought to rest.

b Find the value of e.

c Explain why there must be a third collision between P and Q. [(c) Edexcel Limited 2003]

8 A tennis ball of mass 0.2 kg is moving with velocity $(-10\mathbf{i})\text{ m s}^{-1}$ when it is struck by a tennis racket. Immediately after being struck, the ball has velocity $(15\mathbf{i} + 15\mathbf{j})\text{ m s}^{-1}$. Find

a the magnitude of the impulse exerted by the racket on the ball

b the angle, to the nearest degree, between the vector $\mathbf{i}$ and the impulse exerted by the racket

c the kinetic energy gained by the ball as a result of being struck. [(c) Edexcel Limited 2003]

9 Two small smooth spheres, P and Q, of equal radius, have masses $2m$ and $3m$ respectively. The sphere P is moving with speed $5u$ on a smooth horizontal table when it collides directly with Q, which is at rest on the table. The coefficient of restitution between P and Q is e.

a Show that the speed of Q immediately after the collision is $2(1 + e)u$.

After the collision, Q hits a smooth vertical wall which is at the edge of the table and perpendicular to the direction of motion of Q. The coefficient of restitution between Q and the wall is f, $0 < f \leqslant 1$.

b Show that, when $e = 0.4$, there is a second collision between P and Q.

Given that $e = 0.8$ and there is a second collision between P and Q

c find the range of possible values of f. [(c) Edexcel Limited 2004]

10 A particle P of mass $3m$ is moving with speed $2u$ in a straight line on a smooth horizontal table. The particle P collides with a particle Q of mass $2m$ moving with speed u in the opposite direction to P. The coefficient of restitution between P and Q is e.

a Show that the speed of Q after the collision is $\frac{1}{5}u(9e+4)$.

As a result of the collision, the direction of motion of P is reversed.

b Find the range of possible values of e.

Given that the magnitude of the impulse of P on Q is $\frac{32}{5}mu$

c find the value of e. [(c) Edexcel Limited 2005]

11 Two small spheres A and B have mass $3m$ and $2m$ respectively. They are moving towards each other in opposite directions on a smooth horizontal plane, both with speed $2u$, when they collide directly. As a result of the collision, the direction of motion of B is reversed and its speed is unchanged.

a Find the coefficient of restitution between the spheres.

Subsequently, B collides directly with another small sphere C of mass $5m$ which is at rest. The coefficient of restitution between B and C is $\frac{3}{5}$.

b Show that, after B collides with C, there will be no further collisions between the spheres. [(c) Edexcel Limited 2005]

12 A particle P of mass $2m$ is moving with speed $2u$ in a straight line on a smooth horizontal plane. A particle Q of mass $3m$ is moving with speed u in the same direction as P. The particles collide directly. The coefficient of restitution between P and Q is $\frac{1}{2}$.

a Show that the speed of Q immediately after the collision is $\frac{8}{5}u$.

b Find the total kinetic energy lost in the collision.

After the collision between P and Q, the particle Q collides directly with a particle R of mass m which is at rest on the plane. The coefficient of restitution between Q and R is e.

c Calculate the range of values of e for which there will be a second collision between P and Q. [(c) Edexcel Limited 2008]

13 A uniform ladder AB, of mass m and length $2a$, has one end A on rough horizontal ground. The coefficient of friction between the ladder and the ground is 0.6. The other end B of the ladder rests against a smooth vertical wall.

A builder of mass $10m$ stands at the top of the ladder. To prevent the ladder from slipping, the builder's friend pushes the bottom of the ladder horizontally towards the wall with a force of magnitude P. This force acts in a direction perpendicular to the wall. The ladder rests in equilibrium in a vertical plane perpendicular to the wall and makes an angle α with the horizontal, where $\tan \alpha = \frac{3}{2}$.

a Show that the reaction of the wall on the ladder has magnitude $7mg$.

b Find, in terms of m and g, the range of values of P for which the ladder remains in equilibrium.

14 A straight log AB has weight W and length $2a$. A cable is attached to one end B of the log. The cable lifts the end B off the ground. The end A remains in contact with the ground, which is rough and horizontal. The log is in limiting equilibrium. The log makes an angle α to the horizontal, where $\tan \alpha = \frac{5}{12}$. The cable makes an angle β to the horizontal, as shown in the diagram.

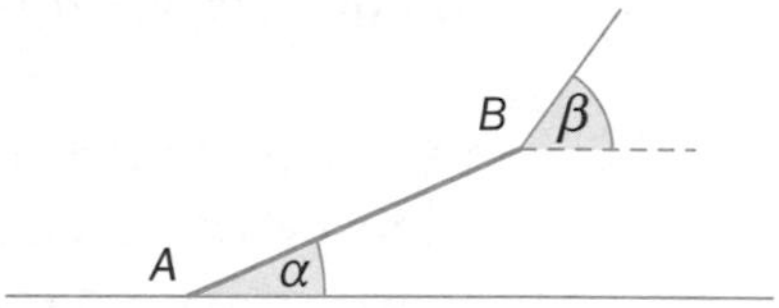

The coefficient of friction between the log and the ground is 0.6. The log is modelled as a uniform rod and the cable as light.

a Show that the normal reaction on the log at A is $\frac{2}{5}W$.

b Find the value of β.

The tension in the cable is kW.

c Find the value of k.

15 A uniform rod AB, of length $8a$ and weight W, is free to rotate in a vertical plane about a smooth pivot at A. One end of a light inextensible string is attached to B. The other end is attached to point C which is vertically above A, with $AC = 6a$. The rod is in equilibrium with AB horizontal, as shown in the diagram.

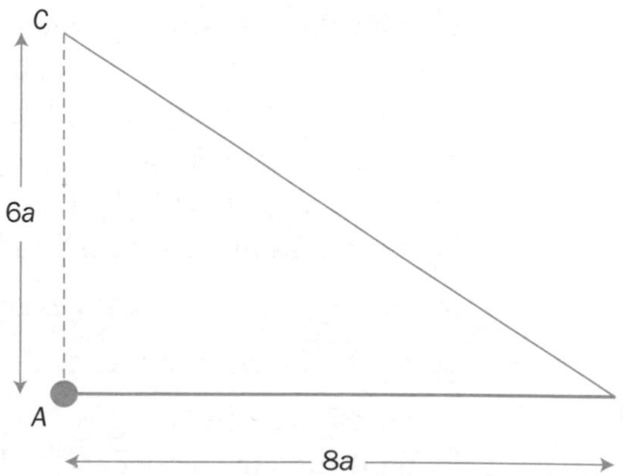

a By taking moments about A, or otherwise, show that the tension in the string is $\frac{5}{6}W$.

b Calculate the magnitude of the horizontal component of the force exerted by the pivot on the rod.

M2

16 A uniform pole AB, of mass 30 kg and length 3 m, is smoothly hinged to a vertical wall at one end A. The pole is held in equilibrium in a horizontal position by a light rod CD. One end C of the rod is fixed to the wall vertically below A. The other end D is freely jointed to the pole so that $\angle ACD = 30°$ and $AD = 0.5$ m, as shown in the diagram. Find

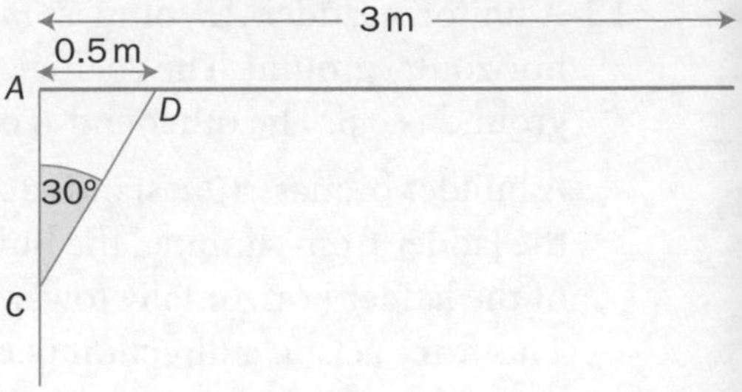

a the thrust in the rod CD

b the magnitude of the force exerted by the wall on the pole at A.

The rod CD is removed and replaced by a longer light rod CM, where M is the midpoint of AB. The rod is freely jointed to the pole at M. The pole AB remains in equilibrium in a horizontal position.

c Show that the force exerted by the wall on the pole at A now acts horizontally.

[(c) Edexcel Limited 2005]

17 A uniform beam AB of mass 2 kg is freely hinged at one end A to a vertical wall. The beam is held in equilibrium in a horizontal position by a rope which is attached to a point C on the beam, where $AC = 0.14$ m. The rope is attached to the point D on the wall vertically above A, where $\angle ACD = 30°$, as shown in the diagram.
The beam is modelled as a uniform rod and the rope as a light inextensible string. The tension in the rope is 63 N. Find

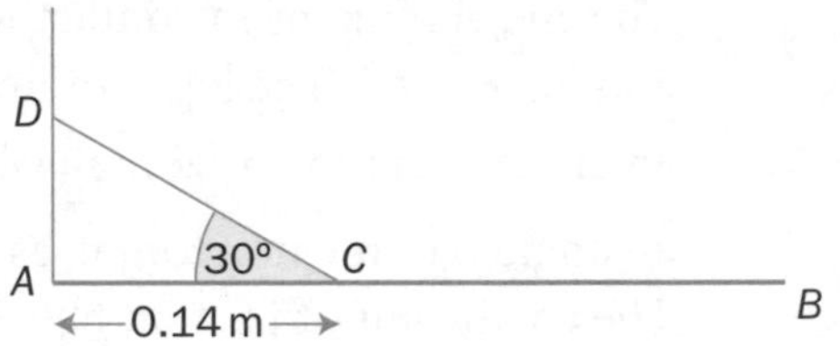

a the length of AB

b the magnitude of the resultant reaction of the hinge on the beam at A.

[(c) Edexcel Limited 2007]

18 A uniform ladder, of weight W and length $2a$, rests in equilibrium with one end A on a smooth horizontal floor and the other end B on a rough vertical wall. The ladder is in a vertical plane perpendicular to the wall. The coefficient of friction between the wall and the ladder is μ. The ladder makes an angle θ with the floor, where $\tan\theta = 2$. A horizontal light inextensible string CD is attached to the ladder at the point C, where $AC = \frac{1}{2}a$. The string is attached to the wall at the point D, with BD vertical, as shown in the diagram.
The tension in the string is $\frac{1}{4}W$.
By modelling the ladder as a rod

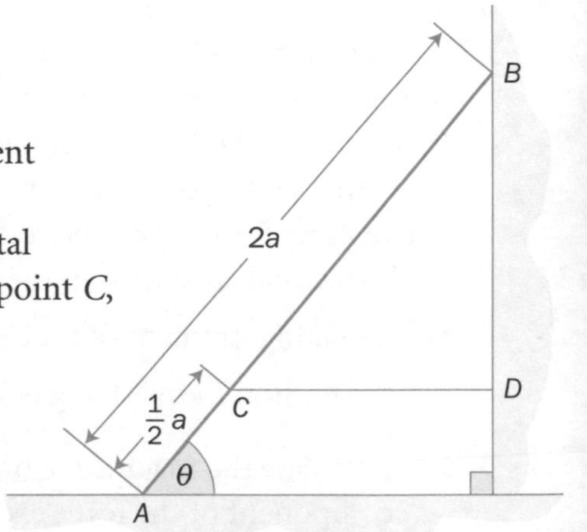

a find the magnitude of the force of the floor on the ladder

b show that $\mu \geqslant \frac{1}{2}$.

c State how you have used the modelling assumption that the ladder is a rod.

[(c) Edexcel Limited 2004]

Answers

Chapter 1

Check in

1 a $1.99\mathbf{i} + 5.61\mathbf{j}$ b 5.95 at 70.4°

2 a $0.5\,\text{m s}^{-2}$ b 8 s

3 a $9x^2 + 2$ b $\frac{5x^{\frac{3}{2}}}{2} - \frac{7x^{-\frac{3}{2}}}{2} - 6x^{-3}$

c $\frac{2}{2x+1}$ d $5e^{5x}$

e $4\cos 4x$ f $3x^2\cos 2x - 2x^3\sin 2x$

4 a $\frac{x^5}{5} + \frac{x^4}{2} + c$ b $\frac{4x^{\frac{3}{2}}}{3} + 10x^{\frac{1}{2}} + c$

Exercise 1.1

1 95.8 m

2 a $\frac{1}{7}$s b 0.1 m

c 3.81° below the horizontal

3 a 1.76 s b 21.6 m c 0.878 s d 3.78 m

4 a 1.77 s b 8.84 m c 0.884 s d 3.83 m

5 a 17.2 m b $15.4\,\text{m s}^{-1}$

c 2.50° below the horizontal

6 a i $30\mathbf{i} + 35.1\mathbf{j}$ ii $60\mathbf{i} + 60.4\mathbf{j}$

b 8.16 s c 245 m

7 30°

8 $31.3\,\text{m s}^{-1}$

9 a $0.021V^2$ b $21.8\,\text{m s}^{-1}$ c 47.7 m

d Ball is a particle, no air resistance, the gym is long enough

10 2.67 s

11 a $20\,\text{m s}^{-1}$, $24.8\,\text{m s}^{-1}$ b 101 m

c $31.9\,\text{m s}^{-1}$ at 51.1° below the horizontal

12 a $7.5\,\text{m s}^{-1}$, $34.8\,\text{m s}^{-1}$ b 64.4 m

13 28.5 m

14 a 2.17 s b 23.0 m

15 $15.3\,\text{m s}^{-1} < u < 18.9\,\text{m s}^{-1}$

16 a $u = 60\,\text{m s}^{-1}$, $v = 9.20\,\text{m s}^{-1}$ b 0.329 m

17 $(14\mathbf{i} + 14\mathbf{j})\,\text{m s}^{-1}$

18 a $54.2\,\text{m s}^{-1}$ at 46.3° to horizontal

b $42.3\,\text{m s}^{-1}$ at 27.6° to horizontal c 5.60 s

19 a $\mathbf{r} = 20t\mathbf{i} + (30t - 4.9t^2)\mathbf{j}$ b $y = 1.5x - 0.01225x^2$

c i 40.4 ii 25.2, 97.3 iii 122 m

20 $V_1 = \frac{3g\sqrt{3}}{4}$, $V_2 = \frac{3g}{2}$

21 $\alpha = \arctan\frac{rq}{p(r-p)}$

22 b $44.1\,\text{m s}^{-1}$ at 29.5° to horizontal

24 b $t = \frac{2V\sin\beta}{g\sin(90+\alpha)} = \frac{2V\sin\beta}{g\cos\alpha}$

Exercise 1.2

1 a $v = 30t - 3t^2$, $x = 15t^2 - t^3$ b $75\,\text{m s}^{-1}$, 250 m

c 500 m d 15 s

2 a −2 m b 6 m

3 a 0 s, 4 s b $\frac{4}{\sqrt{3}}$s

c $-\frac{128}{3\sqrt{3}}$ m d $30\,\text{m s}^{-2}$

4 a $2t^2 + 6t - 8$ m b 1 s

c 4 s before timing started

5 a 1 s, 2 s

b Velocity $-2\,\text{m s}^{-1}$, $3\,\text{m s}^{-1}$ respectively, acceleration $-10\,\text{m s}^{-2}$, $2\,\text{m s}^{-2}$, $8\,\text{m s}^{-2}$ respectively

6 a $t^2 - 5t + 6\,\text{m s}^{-1}$ b 2 s, 3 s c $7\frac{1}{3}$ m, $7\frac{1}{6}$ m

7 a 30 s b 225 m

c Instant aceleration/deceleration (although this would happen if it flew into a constant wind of $2\,\text{m s}^{-2}$)

8 a 2 s b $8 - 10t$ c $-12\,\text{m s}^{-1}$ d 7.2 m

e It gives displacement (height above the ground) not distance travelled.

9 a $\frac{\pi}{2}$s, πs b $2\pi^2\,\text{m s}^{-1}$ c $-4\pi\,\text{m s}^{-2}$, $8\pi\,\text{m s}^{-2}$

10 a $42\,\text{m s}^{-1}$ b 45 m

11 a 6 s b 72 m

12 a e.g. $a = \frac{1}{675}t(t-15)(t-30)$, $v = \frac{1}{2700}t^2(t-30)^2$,

$$x = \frac{1}{13500}t^3(t^2 - 75t + 1500)$$

b $18.75\,\text{m s}^{-1}$ from the given model.

Exercise 1.3

1 a $4\mathbf{i} + 4t\mathbf{j}$, $4\mathbf{j}$

b $2(t-2)\mathbf{i} + t(3t-4)\mathbf{j}$, $2\mathbf{i} + 2(3t-2)\mathbf{j}$

c $\frac{2}{\sqrt{t}}\mathbf{i} - \frac{1}{2t^3}\mathbf{j}$, $-\frac{1}{\sqrt{t^3}}\mathbf{i} + \frac{3}{2t^4}\mathbf{j}$

d $6t(t^2+1)^2\mathbf{i} - \frac{2}{(2t+1)^2}\mathbf{j}$, $6(t^2+1)(5t^2+1)\mathbf{i} + \frac{8}{(2t+1)^3}\mathbf{j}$

e $-2\sin t\,\mathbf{i} + 2\cos t\,\mathbf{j}$, $-2\cos t\,\mathbf{i} - 2\sin t\,\mathbf{j}$

f $e^t\mathbf{i} + \frac{1}{t+1}\mathbf{j}$, $e^t\mathbf{i} - \frac{1}{(t+1)^2}\mathbf{j}$

2 a $\mathbf{r} = \frac{1}{4}t^4\mathbf{i} + t^3\mathbf{j} + \mathbf{c}$, $\mathbf{a} = 3t^2\mathbf{i} + 6t\mathbf{j}$

b $\mathbf{r} = 15t\mathbf{i} + 5t(4-t)\mathbf{j} + \mathbf{c}$, $\mathbf{a} = -10\mathbf{j}$

c $\mathbf{r} = \left(\frac{1}{2}t^2 - \frac{1}{3}t^3\right)\mathbf{i} + \left(\frac{3}{2}t^2 - 5t\right)\mathbf{j} + \mathbf{c}$, $\mathbf{a} = (1-2t)\mathbf{i} + 3\mathbf{j}$

d $\mathbf{r} = (t^2 - 2t^3)\mathbf{i} + (t^3 - t^4)\mathbf{j} + \mathbf{c}$, $\mathbf{a} = (2-12t)\mathbf{i} + (6t - 12t^2)\mathbf{j}$

3 a $\mathbf{v} = 3t^2\mathbf{i} + (2t - 2t^2)\mathbf{j} + \mathbf{c}$, $\mathbf{r} = t^3\mathbf{i} + \left(t^2 - \frac{2}{3}t^3\right)\mathbf{j} + \mathbf{c}t + \mathbf{d}$

b $\mathbf{v} = 2t\mathbf{i} + 2t^2\mathbf{j} + \mathbf{c}$, $\mathbf{r} = t^2\mathbf{i} + \frac{2}{3}t^3\mathbf{j} + \mathbf{c}t + \mathbf{d}$

c $\mathbf{v} = -\frac{2}{t^2}\mathbf{i} + 2\sqrt{t^3}\mathbf{j} + \mathbf{c}$, $\mathbf{r} = \frac{2}{t}\mathbf{i} + \frac{4\sqrt{t^5}}{5}\mathbf{j} + \mathbf{c}t + \mathbf{d}$

d $\mathbf{v} = -6\sin 2t\,\mathbf{i} + 6\cos 2t\,\mathbf{j} + \mathbf{c}$, $\mathbf{r} = 3\cos 2t\,\mathbf{i} + 3\sin 2t\,\mathbf{j} + \mathbf{c}t + \mathbf{d}$

4 a $2(t^2+1)\mathbf{i} + (2t^4 - 1)\mathbf{j}$ b $(9 - 6\cos t)\mathbf{i} + 2(t+1)\mathbf{j}$

5 a $\mathbf{v} = (3t - t^2 + 3)\mathbf{i} + \frac{1}{2}t^2(2 - 3t^2)\mathbf{j}$,

$$\mathbf{r} = \frac{1}{6}(9t^2 - 2t^3 + 18t + 6)\mathbf{i} + \frac{1}{30}(10t^3 - 9t^5 - 60)\mathbf{j}$$

b $\mathbf{v} = 2(1 + \sin 2t)\mathbf{i} + (5 - 4\cos 2t)\mathbf{j}$, $\mathbf{r} = (2t - \cos 2t - 1)\mathbf{i} + (5t - 2\sin 2t + 4)\mathbf{j}$

6 a $3\mathbf{i} + 4(1-t)\mathbf{j}$, $-4\mathbf{j}$ b $5\,\text{m s}^{-1}$ c 1 s

d No, there is always an x-component of velocity.

M2

7 $6.32\,\text{m s}^{-1}$
8 a $2(t^2 - t + 1)\mathbf{i} + (t^3 + 3)\mathbf{j}$ b $4\mathbf{i} + 9\mathbf{j}\,\text{m s}^{-1}$
9 b $28.1°$
10 a $2\mathbf{i} + 2(2 - t)\mathbf{j}$, $2(t + 1)\mathbf{i} + (4t - t^2 - 3)\mathbf{j}$ b 2 s
c 1 s or 3 s
11 $\mathbf{v} = (2t^2 - 2)\mathbf{i} + (12 - 3t)\mathbf{j}$,
$\mathbf{r} = \left(\frac{2}{3}t^3 - 2t + 2\right)\mathbf{i} + \left(3 + 12t - \frac{3}{2}t^2\right)\mathbf{j}$.
When $t = 6$, $\mathbf{v} = 70\mathbf{i} - 6\mathbf{j}$, $\mathbf{r} = 134\mathbf{i} + 21\mathbf{j}$
12 $6\,\text{m s}^{-1}$, $\mathbf{r} = 9\mathbf{i} - 9\mathbf{j}$
13 a $\left(\frac{1}{2}t^2 - 2t\right)\mathbf{i} + t^2\mathbf{j}$ b 4 s c $1\frac{1}{3}$ s
14 a 1 s b 7.16 m
15 a $\mathbf{v} = 2t\mathbf{i} - 4t^3\mathbf{j}$, $\mathbf{r} = (t^2 + 1)\mathbf{i} + (1 - t^4)\mathbf{j}$
b $y = 2x - x^2$, quadratic function, so a parabola.
16 Same when $t = 1$ s, never opposite.
17 a $(3 + 16t)\mathbf{i} + (1 + 24t)\mathbf{j}$ b $16\mathbf{i} + 24\mathbf{j}$
c 2 s d $40\mathbf{i} + 56\mathbf{j}$ away

Review 1

1 a 23.5 m b 78.8 m c $18.0\,\text{m s}^{-1}$
d Ball is a particle, no air resistance
2 a 1.2 s b Clears the wall by 0.044 m
c $10.7\,\text{m s}^{1}$ at 20.6° below horizontal
3 a 4 s b 42.4 c 68.3° below horizontal
4 b $V^2 \sin\theta \cos\theta = 15g$ c 17.2
5 b $\theta > 37.8°$ c 3.50 s
6 a 1.60 s b 0.963 m
c $22.7\,\text{m s}^{-1}$ at 60.3° below horizontal
7 a 40 m b $v = \frac{3t}{5} + \frac{9t^2}{100} - \frac{t^3}{125}$
c $a = \frac{3}{5} + \frac{9t}{50} - \frac{3t^2}{125} = 0\,\text{m s}^{-2}$ when $t = 10$ s d $7\,\text{m s}^{-1}$
8 a $12\,\text{m s}^{-1}$ b $x = 6t^2 - t^3$, 9 m c 32 m, 4 s
d 23 m
9 a $8t - 8e^{-2t}$ b $8 + 16e^{-2t}$
c Acceleration approaches a constant $8\,\text{m s}^{-2}$
10 a $4 + 7t - 2t^2$ c $4t + \frac{7t^2}{2} - \frac{2t^3}{3}$ d $29\frac{1}{3}$ m
11 a $-0.6t$ b $30 - 0.3t^2$ c 10 s d 200 m
12 a $(3t^2 - 3)\mathbf{i} + (2t - 4)\mathbf{j}$ c 2
13 a $4t\mathbf{i} - 8\mathbf{j}$ c $\left(\frac{2}{3}t^3 + 10t\right)\mathbf{i} - 4t^2\mathbf{j}$ d 60 m
14 a 1 m, circle centre O and radius 1 m
b $-2\sin 2t\,\mathbf{i} + 2\cos 2t\,\mathbf{j}$, constant $2\,\text{m s}^{-1}$
15 a $-\mathbf{i} + 3\mathbf{j}$ b $5\,\text{m s}^{-1}$, 126.9° to x-direction
c $4\sqrt{5}$ N at 153.4° to x-direction
16 a 4.20 s b 4.44 s

Chapter 2

Check in

1 $28g$ Nm, $7g$ N

Exercise 2.1

1 1.875 m
2 a 0.9 m b 5
3 a $\left(3\frac{4}{15}, 5\frac{8}{15}\right)$ b (1.3, 3.15) c (1.76, −5.88)
4 a $4\frac{7}{16}\mathbf{i} - 2\frac{3}{8}\mathbf{j}$ b $1\frac{1}{15}\mathbf{i} - \frac{14}{15}\mathbf{j}$
5 3.81 m
6 a 0.9 m b 1.6 m
7 a 0.15 m b 0.5 m
8 (0.8, 1.3)
9 (−13.6, −8.4)
10 (0.24, 0.26)
11 $m = 4$ kg, $DE = 3$ m
12 a (0.16, 0.1) b 0.7
13 b $m_1 = 1$, $m_2 = 2$
14 $m = 2$, (2, 3)

Exercise 2.2

1 a (6.5, 6) b (7, 5) c (6, 5)
d $\left(5, 5 + \frac{16}{3\pi}\right) = (5, 6.70)$ e (1, −1) f (−1, 1)
2 a (5.3, 6.4) b (5.44, 4.46) c (5.27, 3.82)
d (5, 5) e (6.09, 5.17) f (5.07, 5.75)
3 a $\left(2\frac{5}{16}, 1\frac{11}{16}\right)$ m b (1.54, 0.641) m
c (0.923, 0.8) m d (−0.049, 0) m
e (3.20, 1.81) m f (59.7, 67.8) cm
4 (0.578, 0.356) m
5 (41.1, 17.8) cm
7 15.8 cm
8 0.125 cm from O in the opposite direction to the angle bisector of AOB
9 4.38 cm
10 $\frac{2\sqrt{3}}{9}$ m
11 3.23 cm
12 8.75 cm
13 a Symmetry b 0.802 m c 0.391 m
d Because r would need to be > 0.4 m
14 0.304 m
15 a 4.5 b $\frac{9}{80}$ m
16 b Not necessary – no use of the symmetry in the proof.
17 $\frac{47h}{117}$

Exercise 2.3

1 a i 1.5 m ii 1.25 m b 50.2° c 15.9°
2 21.8°
3 a $\left(\frac{2}{5}, \frac{1}{3}\right)$ m b 13.5°
4 a (0.5, 0.15) m b 12.1°
5 32.5°
6 42.0°
7 2.75
8 24.8°
9 18.3 cm
10 a $\left(3\frac{3}{7}, 0\right)$ b 30.3°
11 a 21.8° b 29.7° c It will slide

Review 2

1 2.9 m
2 a 86.9 cm
b 4.72 N down on handle, 11.1 N up on shaft
3 (4.1, 4.7)
4 1.2 m from *AB*, 1.575 m from *AD*
5 a 6 cm from *AB*, 2.5 cm from *AC* b 47.5°
6 a (0.117, 0.133) b 26.6°
7 a 5.79 cm b 0.585 kg
8 30.5°
9 b 8.46 cm

Revision exercise 1

1 a 4 s b $18 - \frac{1}{12}t^3$
2 a 4.02 N b 67**i** + 28**j**
3 b 15 s c 183 m
4 a 1.05 s c 11.7 m s^{-1}
d *B* (and *T*) are not particles.
5 6 s
6 b 3.5 s
7 a 6 cm b 22.6°
8 b $\frac{3}{55}$
9 a 6.86 cm b 32.1°
10 b $\left(\frac{16}{3}, \frac{70}{27}\right)$ c 25.9°
11 a $\left(4\frac{1}{2}, \frac{2}{3}\right)$ c $\frac{4}{9}$ d 83.7°

Chapter 3

Check in

1 $v = 2\sqrt{gh}$
2 a 13**i** + 43**j** b $\sqrt{89}$
3 $v = u + at,\ s = \frac{1}{2}(u+v)t,\ s = ut + \frac{1}{2}at^2,$
$s = vt - \frac{1}{2}at^2, v^2 = u^2 + 2as$

Exercise 3.1

1 a 3600 J b 2950 J
2 5000 J
3 13 700 J
4 a 1760 J b 6760 J
5 35 m
6 a 221 J b $0.25gn(n-1)$ J
7 1060 J
8 a 0.5 m s^{-2} b 33 N c 528 J
9 2880 J
10 a 2250 J b 3720 J
11 a 268 J b 563 J
12 b 1300 J
13 2780 J
14 a 1020 m, 102 kJ b 1250 m, 108 kJ
c 557 m, 52.4 kJ
15 a 1910 J b 2160 J
16 b $v = 2\sqrt{gx\sin\alpha}$

Exercise 3.2

1 a $3gh$ J b 1176 J c 40 m
2 a 64 J b 16 J c 48 J d 96 m
3 15.3 m s^{-1}, mass not needed.
4 a $16m$ J b $16m$ J c 0.163
5 a 8.85 m s^{-1} b i 530 J ii $88\frac{1}{3}$ N
6 20 m
7 a 18.4 m s^{-1} b 50.7 m c 19.0 m s^{-1}
8 a 12.5 m s^{-1} b 12.1 m s^{-1}
9 4.6 N
10 4.70 m s^{-1}
11 a 3.43 m s^{-1}
b There would be a jerk (sudden change) as the string went taut.
12 a 6.66 m s^{-1} b 3.05 m s^{-1}
13 4.235 m s^{-1}
14 a $\sqrt{2ga\cos\theta}$ b $\sqrt{2ga}$
15 $\sqrt{\frac{2ga}{3}}$
16 $\sqrt{2}:1$

Exercise 3.3

1 a 39.2 W b $6\frac{8}{15}$ s
2 2250 W
3 a $1998a$ J b 0.25 m s^{-1}
4 500 N
5 25.1 W
6 25.0 W
7 2 MW
8 14 500 W
9 a 10 m s^{-1} b 0.625 m s^{-2}
10 a 200 N b 8.45 m s^{-1} or 30.4 km h^{-1}
11 20.6 m s^{-1}
12 30 m s^{-1} or 108 km h^{-1}
13 a −0.030 m s^{-2}, 9.05 m s^{-1} b 0.255 m s^{-2}
14 837 m
15 a 11 200 W b i 0.773 m s^{-2} ii 284 N
16 0.221 m s^{-2}
17 a 2520 N b 22.3 m s^{-1}
19 5.2 m s^{-1}

Review 3

1 a 1180 J b 995 J
2 a 43.2 J b 43.2 J c 1.44 m
3 a 0.8 J b 0.816 m
c 12.6 J d 15.8 m s^{-2}
4 a 10.8 m s^{-1} b 2.93 N
5 a 288 J b 288 J c 0.367
6 a 1200 N b 0.632 m s^{-2}
c 18.5 m s^{-1} or 66.6 km h^{-1}
7 a 112 kW b 2.8 m s^{-1} c 57.6 m s^{-1}
8 a 200 W b 25.2°
9 a 2000 N b 0.98 m s^{-2} c 58.8 m s^{-1}

Chapter 4

Check in

1 $u = 4, v = -1$
2 24 J
3 $v = u + at$, $s = \frac{1}{2}(u + v)t$, $s = ut + \frac{1}{2}at^2$, $s = vt - \frac{1}{2}at^2$, $v^2 = u^2 + 2as$

Exercise 4.1

1 $(10\mathbf{i} + 20\mathbf{j})$ Ns
2 a i $3.25\,\text{m s}^{-1}$
ii $-1.25\,\text{m s}^{-1}$ (travelling in *B*'s direction)
b i 0.375 J ii 18.375 J
3 a $2\,\text{m s}^{-1}$ b 12 Ns c 48 J
d *A* would have to pass through *B*.
4 a 6 b $1.8\,\text{m s}^{-1}$

Exercise 4.2

1 a 0.75 b 0.5 c $\frac{5}{9}$
2 a $2.51\,\text{m s}^{-1}, 7.31\,\text{m s}^{-1}$ b $2.64\,\text{m s}^{-1}, 4.24\,\text{m s}^{-1}$
c $0.91\,\text{m s}^{-1}, 5.71\,\text{m s}^{-1}$ d $-2.71\,\text{m s}^{-1}, -0.91\,\text{m s}^{-1}$
3 a $1\,\text{m s}^{-1}, 2\,\text{m s}^{-1}$ b $\frac{2}{11}$
4 a $5.6\,\text{m s}^{-1}$ b 0.2
5 a 0.6 b 15 J
6 $-7.38\,\text{m s}^{-1}, 4.62\,\text{m s}^{-1}$
7 mu^2 J
8 b Yes
9 a i $4\,\text{m s}^{-1}, 9\,\text{m s}^{-1}$ ii $6\,\text{m s}^{-1}$ iii 60 J
b Final momentum = initial momentum independent of *e*
10 $k = \frac{5}{16}$ or 4.25
11 2.68 m
12 a 8 s b 6

Exercise 4.3

1 a $\frac{1}{3}$ b $5.6\,\text{m s}^{-1}$ c $15\,\text{m s}^{-1}$
2 a 9 s b 15 J
3 a $8.4\,\text{m s}^{-1}$ b $2.8\,\text{m s}^{-1}$ c $\frac{1}{3}$ d 4.5 m
4 b $1\,\text{m s}^{-1}$
5 a $7.5\,\text{m s}^{-1}, 10\,\text{m s}^{-1}$ b $0\,\text{m s}^{-1}, 7.5\,\text{m s}^{-1}$
c A 3rd collision which brings *B* to rest
6 a i $-2.2\,\text{m s}^{-1}, -0.96\,\text{m s}^{-1}$ ii $-2\,\text{m s}^{-1}, -0.4\,\text{m s}^{-1}$
b i 53.3 J ii 55.6 J
7 $16\sqrt{g}$
8 b $\frac{2d(e_1 - e_2)}{ue_1e_2}$

Exercise 4.4

1 $2.03\,\text{m s}^{-1}, 2.34\,\text{m s}^{-1}, 5.63\,\text{m s}^{-1}$
3 b 23.5 J
4 a $4\,\text{m s}^{-1}$ b 6 Ns c $2\,\text{m s}^{-1}$
5 a $\frac{u(1-e)}{2}, \frac{u(1-e^2)}{4}, \frac{u(1+e)^2}{4}$

Review 4

1 a 12 s b 30 m
2 a $6\,\text{m s}^{-1}$ b 0.25
3 a *P* $4\,\text{m s}^{-1}$, *Q* $1\,\text{m s}^{-1}$, both in *Q*'s original direction
b 10 Ns
c 5 J
4 a $6\,\text{m s}^{-1}, 4\,\text{m s}^{-1}$ b $5\frac{5}{7}$ m
5 a 2.5 m b 1.225 m c $9\frac{41}{51}$ m
6 a i $2.4\,\text{m s}^{-1}$ ii $-2.7\,\text{m s}^{-1}, -1.5\,\text{m s}^{-1}$
iii $-1.8\,\text{m s}^{-1}$
b After the spheres collide, *B* is at rest and *A* has speed $2.4\,\text{m s}^{-1}$. Final common speed = $0.6\,\text{m s}^{-1}$. Loss of KE = 23.3 J.
7 a $\frac{V}{3}$ b $\frac{V}{12}, \frac{V}{4}$
8 b $12.25\,\text{m s}^{-1}, 8.75\,\text{m s}^{-1}$

Chapter 5

Check in

1 a $T\cos 30° - W\sin 20°$
b $R + T\sin 30° - W\cos 20°$
2 $6T\sin 40° - 3W$
3 a $F = 0.3R$ b $F \leqslant 0.3R$

Exercise 5.1

1 a 63.7 N b 73.5 N c 127 N
2 a 23.4 N
b 45.6 N at 59.2° above horizontal towards wall
3 a 51.7 N, 38.4 N b 1.63 m
4 b $AD = \frac{a}{3}$, $P = \frac{2W\sqrt{3}}{3}$
5 21.2 N, 61.2 N, 42.4 N
6 a 34.6 N b 2.33 m
7 63.4°
8 61.9°
9 29.7 N, 0.152
10 $0.2W, 0.6W$
11 a $\frac{a}{\sqrt{15}}$ b $3.04W$, 70.8° below horizontal
13 a 71.9 N b 92.9 N, 50.7° to vertical
14 a 5.94 m b 19.2

Review 5

1 a 92.4 N b 61.1 N at 40.9° below horizontal
2 a 199 N, 374 N b 4.13 m
3 a 28.3 N b 49 N
4 $\frac{1}{2\sqrt{3}}$
5 a 113 N b 0.348
6 a $\frac{20}{\pi}$ N b 21.0 N at 72.3° above horizontal
8 $1.625a$
9 b 2.08 N c 6.54 N
10 a Smooth ring, so reaction and therefore tension must be perpendicular to the rod
c $1.29W$

M2

Revision exercise 2

1 **a** 0.8 **b** $81\frac{2}{3}$
c Resistance may vary with speed.

2 **a** 324 W **b** $9.33\,\text{m s}^{-1}$
c 32 N **d** $0.591\,\text{m s}^{-2}$

3 **a** 50 J **b** 0.319

4 **a** $\frac{7mgh}{5}$ **b** $\frac{3gh}{5}$

5 **a** 304 J **b** 0.674

6 **b** 102.4

7 **b** $\frac{25}{32}$
c Q will bounce off the wall again and hit P for a third time.

8 **a** 5.83 Ns **b** 31° **c** 35 J

9 **c** $f > \frac{1}{9}$

10 **b** $e > \frac{2}{3}$ **c** $\frac{7}{9}$

11 **a** $\frac{2}{3}$

12 **b** $0.45\,mu^2$ **c** $e > 0.25$

13 **b** $0.4\,mg \leqslant P \leqslant 13.6\,mg$

14 **b** 68.2° **c** 0.646

15 **b** $\frac{2}{3}W$

16 **a** 1020 N **b** 778 N

17 **a** 45 cm **b** 55.8 N

18 **a** $\frac{7W}{8}$ **c** Negligible thickness, does not bend

M2

Glossary

Motion in two dimensions A particle moving in 2D has displacement $\mathbf{r} = x\mathbf{i} + y\mathbf{j}$, velocity $\mathbf{v} = v_x\mathbf{i} + v_y\mathbf{j}$ and acceleration $\mathbf{a} = a_x\mathbf{i} + a_y\mathbf{j}$.

For uniform acceleration, the displacement $\mathbf{r}$, initial velocity $\mathbf{u}$, final velocity $\mathbf{v}$, acceleration $\mathbf{a}$ and time t are connected by:

$$\mathbf{v} = \mathbf{u} + \mathbf{a}t$$
$$\mathbf{r} = \frac{1}{2}(\mathbf{u} + \mathbf{v})t$$
$$\mathbf{r} = \mathbf{u}t + \frac{1}{2}\mathbf{a}t^2$$
$$\mathbf{r} = \mathbf{v}t - \frac{1}{2}\mathbf{a}t^2$$

In addition $v_x{}^2 = u_x{}^2 + 2a_x x$ and $v_y{}^2 = u_y{}^2 + 2a_y y$

A projectile is modelled as a particle moving under gravity with no air resistance. If initial speed is u and angle of projection is α, the particle obeys the equations for uniform acceleration with $\mathbf{u} = u\cos\alpha\,\mathbf{i} + u\sin\alpha\,\mathbf{j}$ and $\mathbf{a} = -g\,\mathbf{j}$

The maximum height for a projectile occurs when $v_y = 0$.

The range of a projectile is the distance between its point of projection and the point at which it returns to the same horizontal plane. For a given value of u, elevations of α and $(90° - \alpha)$ give the same range. Maximum range is achieved when $\alpha = 45°$.

For non-uniform acceleration the displacement, velocity and acceleration are functions of time.

Velocity is the rate of change of displacement:

$$v = \frac{dx}{dt}, \text{ also written as } v = \dot{x}$$
$$\mathbf{v} = \frac{d\mathbf{r}}{dt} = \dot{\mathbf{r}} = f'(t)\mathbf{i} + g'(t)\mathbf{j}$$

Acceleration is the rate of change of velocity:

$$a = \frac{dv}{dt} = \frac{d^2x}{dt^2}, \text{ also written as } a = \dot{v} = \ddot{x}$$
$$\mathbf{a} = \frac{d\mathbf{v}}{dt} = \frac{d^2\mathbf{r}}{dt^2} = \ddot{\mathbf{r}} = f''(t)\mathbf{i} + g''(t)\mathbf{j}$$

You integrate acceleration to obtain velocity:

$$v = \int a\,dt$$
$$\mathbf{v} = \int \mathbf{a}\,dt$$

You integrate velocity to obtain displacement:

$$x = \int v\,dt$$
$$\mathbf{r} = \int \mathbf{v}\,dt$$

Differentiating/integrating a vector involves differentiating/integrating its components separately.

Centre of mass The centre of mass of an object is the point through which its weight acts when it is in a uniform gravitation field. For particles of mass $m_1, m_2, \ldots, m_n$ placed at points with coordinates $(x_1, y_1), (x_2, y_2), \ldots, (x_n, y_n)$, the centre of mass is $G(\bar{x}, \bar{y})$, where

$$\bar{x} = \frac{m_1x_1 + m_2x_2 + \cdots + m_nx_n}{m_1 + m_2 + \cdots + m_n} = \frac{\sum_{i=1}^{n} m_i x_i}{\sum_{i=1}^{n} m_i} \quad \text{and}$$

$$\bar{y} = \frac{m_1y_1 + m_2y_2 + \cdots + m_ny_n}{m_1 + m_2 + \cdots + m_n} = \frac{\sum_{i=1}^{n} m_i y_i}{\sum_{i=1}^{n} m_i}$$

For standard shapes you find G using symmetry or the formulae provided. For composite bodies treat each component as a particle at its centre of mass.

When a body is suspended in equilibrium from a single point P, its centre of mass is vertically below P.

An object resting on a surface will topple if a vertical line through its centre of mass does not pass through its region of contact with the surface.

Work A force does work if its point of application is displaced in the direction of the force.

If the point of application of F N is displaced s m in the direction of F, then

work done by $F = Fs$ joules (J)

If the displacement is at an angle θ to the direction of F, then

work done by $F = Fs\cos\theta$ J

Work done by friction is negative. You usually refer to work done against friction (that is, the work done by the system in overcoming friction).

If an object is raised, its weight does negative work. You usually refer to work done against gravity.

Energy An object has (mechanical) energy if it has the capacity to do work.

An object in a gravitational field has gravitational potential energy (GPE). You choose a zero level from which to measure GPE.

For an object of mass m kg at a height h m from the zero level

GPE $= mgh$ **J**

An object which is moving has kinetic energy (KE).

An object of mass m kg travelling at v m s^{-1} has

KE $= \frac{1}{2}mv^2$ **J**

The principle of conservation of mechanical energy: The total mechanical energy of a system remains constant provided no external work is done and there arc no sudden changes (collisions or jerks) in the motion of the system.

The work-energy principle: The total work done on or by a system equals the change in mechanical energy of the system.

Power Power is the rate at which work is done.
The SI unit of power is the watt (W), where $1\,\mathbf{W} = 1\,\mathbf{J\,s^{-1}}$
For a vehicle travelling at $v\,\text{m s}^{-1}$ under a driving force F N
Power = Fv W

Collisions An object of mass m kg moving with velocity $\mathbf{v}\,\text{m s}^{-1}$ has **momentum** = $m\mathbf{v}$ N s
When objects collide, each receives an **impulse I**.
Impulse = change of momentum: $\mathbf{I} = m\mathbf{v} - m\mathbf{u}$

If no external force acts on a system, the total momentum of the system remains constant.
This is the principle of conservation of momentum.

For collisions between objects which are free to move, momentum is conserved. For a collision involving a fixed object (e.g. a particle hitting a wall), momentum is not conserved.

All collisions obey Newton's experimental law of restitution:

$$\frac{\text{separation speed}}{\text{approach speed}} = e \text{ (a constant)}$$

e is the coefficient of restitution.
If $e = 0$ the impact is inelastic. No rebounding occurs.
If $e = 1$ the impact is perfectly elastic.
Unless an impact is perfectly elastic there is a loss of kinetic energy.

Equilibrium A system of coplanar forces is in equilibrium if

- The resultant force in each of two directions is zero.
- The total moment about any chosen point is zero.

There are three strategies for obtaining equations leading to the solution of a problem.

1. Resolve in two (usually perpendicular) directions and take moments about one chosen point. Equate the totals in each case to zero.
2. Resolve in one direction and take moments about two points. Equate the totals in each case to zero.
3. Take moments about three points. Equate the totals in each case to zero.

When a system of three non-parallel forces is in equilibrium, the lines of action of the forces must all pass through a single point.

Formulae

M2

The following formulae will be given to you in the exam formulae booklet.
You may also require those formulae listed under Mechanics M1 and Core Mathematics C1, C2 and C3.

Centres of mass

For uniform bodies:

Triangular lamina: $\frac{2}{3}$ along median from vertex

Circular arc, radius r, angle at centre 2α: $\frac{r \sin \alpha}{\alpha}$ from centre

Sector of circle, radius r, angle at centre 2α: $\frac{2r \sin \alpha}{3\alpha}$ from centre

Index